Mobile Library Services

Best Practices

Edited by
Charles Harmon
Michael Messina

THE SCARECROW PRESS, INC.
Lanham • Boulder • New York • Toronto • Plymouth, UK
2013

Published by Scarecrow Press, Inc.
A wholly owned subsidary of The Rowman & Littlefield Publishing Group, Inc.
4501 Forbes Boulevard, Suite 200, Lanham, Maryland 20706
www.rowman.com

10 Thornbury Road, Plymouth PL6 7PP, United Kingdom

British Library Cataloguing in Publication Information Available

Library of Congress Cataloging-in-Publication Data

Mobile library services : best practices / edited by Charles Harmon, Michael Messina.
  p. cm.
 Includes bibliographical references and index.
 ISBN 978-0-8108-8752-7 (pbk. : alk. paper) — ISBN 978-0-8108-8753-4 (ebook)
 1. Mobile communication systems—Library applications. 2. Mobile communication
systems—Library applications—Case studies. 3. Public services (Libraries)—
Technological innovations. I. Harmon, Charles, 1960– II. Messina, Michael.
 Z680.5.M63 2013
 025.0422—dc23   2012043441

Contents

Introduction

SCOTT LA COUNTE

Author of *Going Mobile: Developing Apps for Your Library Using Basic HTML Programming*

June 29, 2007. Nothing about that date seems important—at first glance. But that's the day that changed everything about information as we know it. That is the day the iPhone was introduced.

The iPhone was a phone that could connect you to the Internet from anywhere in the world. In essence, the iPhone meant that nearly all the information in the world could be stored in your pocket . . . but that's not exactly innovating—that's an idea that had been around for over 10 years. Nokia smartphones were really the first to capitalize on the idea of mobile computing, and that was in the nineties. So what made the iPhone so innovating? In a word: apps.

Apps had been around for quite some time as well, but Apple did something that no other company had—it made people want them. It courted developers and turned them into millionaires. It started a postmodern gold rush. Developers were literally making millions of dollars, 99 cents at a time. Before the end of the year, developers from every continent were capitalizing on apps. By the end of the year, Google already had engineers working on their own mobile operating system. In short, everybody wanted in on what Apple had discovered.

Everyone, that is, except libraries. Even today, few libraries have gone mobile. Libraries used to be at the forefront of any technology boom—they were among the first to have computers, Internet, and even e-books (before people, by and large, had even heard of e-readers—most librarians probably remember Stephen King making e-book headlines with *Riding the Bullet* nearly 10 years before Amazon released the first Kindle). Budget cuts and untrained staff have plagued the digital resources of libraries big and small—but that's only an excuse and not a reason for the lack of innovation. In reality, mobile computing can cost the library next to nothing. There are plenty of free and cheap resources available—you just have to know where to look.

This book will show you how your library can create apps, but it will also show something more important: It will show you that you don't have to build apps—apps are just one small part of mobile computing. This book will help you get the library off the desktop and into the palm of your patrons' hands. Most important, this book will help your library stay relevant even on a shoestring budget.

Before we continue on, there are two things to consider: First, what is mobile computing? Second, why does it matter?

Regarding the former, mobile computing can be several things. It can refer to resources that the library can use on feature phones (i.e., the phone that most Americans still use—the basic bar phones or flip phones are considered *feature phones*). Mobile computing can also refer to smartphones (e.g., Android, iPhone, Blackberry, Symbian, Windows phone), and it can refer to tablets. The reality is that mobile devices can refer to essentially any device that someone uses on the go. Unfortunately, unlike computers—where you can get a website to run smoothly regardless of the browser or operating system—mobile computing rarely works this seamlessly.

It's important to identify how your library wants to implement mobile computing. Ideally, you will want to execute a strategy that is useful to all devices. The reality is, however, this can't always happen. Many mobile solutions, such as a "text a librarian" option for patrons, will let libraries on a shoestring budget have a solution that is appealing to nearly all of their mobile users—these are the kind of solutions that you want to look into before considering something a little more complex and device specific.

As you begin to explore what steps to take, remember to keep it social. Mobile computing isn't a small undertaking. Group collaboration is a key strategy. You will need input from young and old users alike. With so many directions to go, working in groups will help you stay focused and ensure that you don't get offtrack or lose sight of your goal.

It sounds like a lot of work, but don't worry—it doesn't have to be—which brings us to the second inquiry: Why does mobile computing matter? It's easy to raise that question considering the amount of people lacking smartphone devices. Before you ponder it to deeply, look around—how many teens would rather text a question than ask one out loud? How many people now prefer eBooks to real books? How many people have their phones glued to their hands 24/7? In short, every year the number of people relying on mobile devices goes up. There will always be people who want reference the old-fashioned way, and we can't forget about them—but we also can't neglect people who want reference on the go.

It's important to consider as well the number of people who are abandoning computers altogether in favor of tablets that they find easier to use. Tablets are relevant for the mere fact that there is almost no learning curve. They are perfect for people who only want e-mail, social networking, and light Internet browsing. These people are expecting resources that are easy to use, and libraries need to find them. They don't want pages of instructions—they want a two- or three-step process. This means that now, more than ever, libraries need to look at the user interface of any resource and question its usability: Is it easy to use? Can the website be used by someone with a touch screen? Do the colors show up well inside and outdoors? Is the text the right size? Look at Apple's app store—reviews are critical! If you rush out the gates just to do something mobile, your patrons will likely respond unfavorably. When considering mobile computing, remember that first impression is everything. Patrons have seen the power of their phone/tablet, and they don't want second best.

As you read this book, decide what your budget is, and remember that there really is something for everyone. This book will show you free and paid resources alike; it will show you what to do if you have lots of time and what to do if you have only a little. If there is a larger budget, remember that there are plenty of contractors online—check out elance.com and odesk.com. On these sites, you can find mobile developers who can build and maintain mobile resources for as little as $1,000. Paid contractors often require more direction-like communications than do subscription-type mobile services, but the result is something that can be more relevant to your patrons.

Mobile computing is the future. It is not only what people want; it is what they will expect. It is not time to consider the possibility of going mobile—that time has come and gone. Mobile computing is here and will not go away. The urgency to offer mobile solutions grows more every day. It is an exciting time to offer reference, and librarians are at the forefront of how people will get information. For the first time since the Internet was first adopted by the mainstream, libraries get to offer something new and different—to pave the way for the next generation of librarians. Now is the time for libraries to do what they do best: adapt, enhance, and innovate.

1

A Student–Library Collaboration to Create CULite: An iPhone App for the Cornell University Library

Matthew Connolly and Tony Cosgrave

Cornell University Library

BEGINNING THE PROJECT

In 2007, the Cornell University Library (CUL; 2012a) established a set of high-level priorities, one of which was to "expedite access to scholarly resources at the point and place of need." To further this goal, the library established a group to find new ways to bring library resources and services to its users outside the library. In 2009, a number of factors motivated this group (Library Outside the Library [LOL]) to begin investigating the possibility of creating a mobile presence for the library's patrons. Library staff regularly saw students using handheld devices at public service desks. Students were coming up to the reference desks with call numbers on their screens and asking how to locate them. Many library staff members had handheld devices and were actively exploring a growing collection of mobile interfaces for information sources and services. At the same time, various library-centric e-mail lists, blogs, and literature were producing more and more content regarding mobile activities in libraries (Chapman, 2009; Kraft, 2009; Read, 2009; Sierra, Ryan, & Wust, 2007). It was clearly time for CUL to start considering mobile access to its resources and services.

The LOL group began its study by looking at access data from its server logs. We saw a dramatic increase in the number of mobile devices accessing library servers during one semester (Kraft, 2009):

September 1–21, 2008: 138 views from iPod/iPhone, 3 views from other mobile devices

February 1–21, 2009: 5,844 views from iPod/iPhone, 4,676 views from other mobile devices

1

We also talked with representatives of the Student Library Advisory Council. This group of approximately 20 students, graduates and undergraduates, meets once a month to provide input on library issues. We showed them examples of mobile library projects from other institutions, including the New York Public Library and North Carolina State University Library. After seeing what these libraries had done, the students encouraged us to develop mobile access to some of CUL's resources and services. As a result of this endorsement from the advisory council and our research, LOL initiated a project in the fall of 2009 to develop an iPhone/iPod touch application that could offer a subset of the library's online resources and services. In line with CUL's strategic goals and priorities, the chief goal of the project was to reach out to our patrons and offer them access to the library from a new platform that was clearly increasing in popularity.

BUILDING THE APP

Working with Students

Each fall semester, the professor of a computer science class at Cornell (CS 5150: Software Engineering) puts out a call for project proposals. Over the course of the semester, the CS 5150 students work in teams to design and implement real-world software projects for clients at the university. In the past, CUL had used student teams from the course to complete projects and had been quite satisfied with their work. Library developers did not have any experience with iPhone application development, and we were concerned about spending staff time to develop a service that would privilege just one category of users. We were aware that other departments of the university were creating iPhone apps and that iPhones and iPods were very popular with students, faculty, administration, and staff on campus. We thus submitted a proposal to the students in CS 5150 to create an app for CUL. Our proposal included a list of some desired resources and services that LOL wanted the app to include, but we encouraged the students to make suggestions of their own. A group of six students met with us to discuss our proposal and subsequently chose us as their project for the semester.

The whole process was handled by the students in a very professional manner. They treated us as real-world clients. They wrote and presented a formal prospectus. LOL members met with them every week to hear about the team's progress, discuss technical issues, and provide feedback. Several times throughout the semester, the students, their professor, and library staff met, and the students gave formal presentations detailing their progress and discussing next steps. Since the students were working primarily on a class project, we didn't have to concern with the material aspects of the development process. We didn't have to order equipment or supplies for them but, rather, simply provide technical assistance and advice. The students did most of their development using their personal Mac laptops, running XCode for programming and their personal iPhones or iPod touches for testing. They set up an internal team wiki to keep track of their project, but library staff did not use or have access to it.

For the students, the most difficult technical aspects of interacting with library data involved connecting to our Voyager catalog to do searches, display patron account information, and list item availability. Two library developers (who also served on the LOL team) acted as technical liaisons for the students, providing information about Voyager application programming interfaces, library systems, and other technical aspects of the system. One of the developers attended meetings with the students regularly, while the other answered questions through e-mail.

Usability Testing

As the students neared completion of a working prototype, we began considering ways to test the interface and functionality before finalizing the app. CUL is fortunate enough to have a dedicated usability committee that handles user testing of library's websites and systems. Although the usability group had never tested anything like an iPhone app before, it readily agreed to try.

The iPhone environment posed some unique challenges to the usability group's standard methodology. Most of its usability tests to date involved setting up a volunteer with a computer running the website or system in question and having him or her use the system to complete a number of tasks (e.g., download a full-text version of the article with the citation x). As the volunteer worked through the tasks, a member of the usability group would take notes, and screen-recording software would capture the test sessions for later analysis.

With no way to output the phone's display to a larger screen, it was difficult for the observer to see what was going on as the volunteer used the app. Recording the session on the phone was out of the question. However, low-tech solutions worked well: With one observer holding a video camera aimed at the iPhone screen and another peering over the volunteer's shoulder to take notes, the tests were recorded successfully. The usability team's report noted some issues with the user interface and functionality, but they were relatively minor. We used this feedback to adjust our interface and system design, but overall development continued uninterrupted.

Costs to the Library

Since we were able to use the student team for the brunt of the work, library resources required for the development effort were minimal. Four representatives from LOL were involved in the project as coordinators and technical consultants from CUL; these four met with the students as needed to continue the project. This time had already been allotted for LOL work and didn't require additional budgeting. Except for advertising and promotion, no money was spent to create the app. We did, however, splurge a bit by taking the students out for dessert to celebrate a successful end to the project!

LOL is an incubator of new efforts and ideas, and its role in the project was to initiate and oversee the development effort. Once the development was complete and the app was ready to launch, it was time to hand off the ownership and maintenance to someone else. DLIT—CUL's Digital Library and Information Technologies group—agreed to support the app, with the understanding that a student worker with developer experience would be hired to handle the lion's share of the work. One of the LOL developers who had consulted with the student team would supervise further student development.

DESCRIPTION OF THE APP

The final product was a working iPhone/iPod touch application that includes the following features:

- Search access to the library catalog
- Access to patron accounts
- Listing of library hours and current open/closed status
- Contact information for CUL unit libraries
- Interactive campus/library map
- "Ask a Librarian" service point (enabling users to consult reference librarians through e-mail, SMS [text], or phone)
- Links to mobile-friendly websites and resources

Figure 1.1 shows several screens from the app.

The catalog search enables users to perform searches in CUL's Voyager OPAC. The search feature allows searches by title, author, or keyword. Search results include item availability and basic bibliographic information. Users can also recall an item. The patron account feature retrieves information about the user's charged items, item requests, and any incurred fines. This feature also allows users to renew items and cancel pending requests.

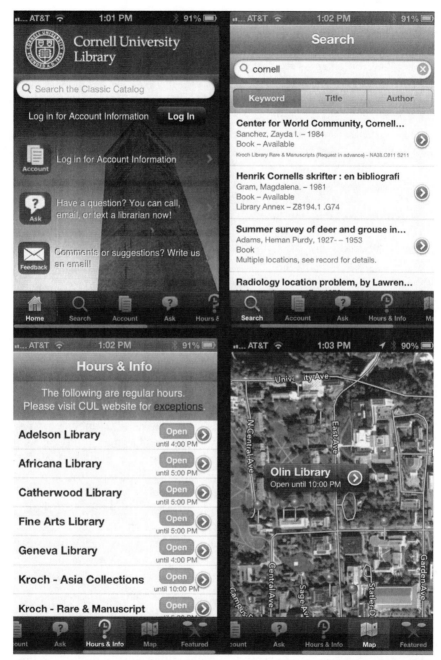

Figure 1.1. Four views of the Cornell University Library iPhone app. Clockwise from upper left: home screen, catalog search, hours status, map showing library locations and hours.

The iPhone development coincided with CUL's transition to Voyager 7, which provided application programming interfaces for accessing Voyager's search functionality and patron records that were needed to enable these features. These interfaces consist of a set of XML web services designed for use in alternative OPACs.

The "Ask a Librarian" service provides contact information for each of CUL's libraries. On an iPhone, the phone, e-mail, and SMS (texting) links will activate the appropriate iPhone apps to place calls or send e-mail and SMS messages to the selected library. On an iPod touch, which does not have phone or SMS capability, the app lists the relevant contact information.

The library hours and map page displays current library hours and a Google map showing which libraries are open and which are closed. Users can find the nearest open library and contact the public service desks there.

The featured links provide a selected list of mobile CUL sites. These include CUL's presence on Flickr and Facebook, mobile-friendly interfaces to licensed databases and reference works, and free mobile websites valuable for research and learning.

LAUNCH AND REACTION

Advertising and Promotion

The library advertised the finished app aggressively. The LOL group worked with the CUL communications department to develop a marketing plan and create a loose brand name: "CULite." CULite services included a new mobile-friendly version of the library website as well as the new iPhone app. Together, we designed posters and tabletop ads that were distributed to all 20 unit libraries. Communications staff wrote an article for the campus news service, and reporters for the student newspaper interviewed numerous people from the project and wrote a story about it (Glazer, 2010; Merola, 2010). We also promoted the app on the library's website. The most creative approach was the production of a whimsical video posted to YouTube.[1] When a new version of the app was later released with an improved log-in system, another university blog featured it in a post.[2]

Distribution

Having a finished iPhone app was useless without a means of distributing it to the public. There are two ways of distributing iPhone applications. The most common by far is to submit the app to Apple's iTunes Store as a free or paid download. There is a second method intended for enterprise app deployment.

This method allows an entity such as a university to sign an app with a special code and distribute it to members of that organization. Patrons downloading the app must have a matching code installed into their copy of iTunes to run the app on their device.

Our ultimate goal was to have the CUL app appear in the iTunes Store so that it could reach library patrons beyond the university. App distribution was new to everyone at CUL, though. Furthermore, the submission and approval process for getting an app to appear in the store could be notoriously slow and mysterious. As a stopgap measure, our initial approach was to use enterprise distribution to make the app available to students and faculty at Cornell. We turned to Cornell Information Technologies (CIT), the central information technology division of the university. CIT was already experimenting with app development and distribution and had a rudimentary central repository of apps that had been developed by different schools or units within the university. There was a single website where these apps were listed, with descriptions and download links, along with instructions for installing the provisioning codes necessary to use them. Once CIT had posted information and a link for our library app, we announced its release to the campus.

At the same time, we were working with a representative at CIT on the iTunes Store submission process. CIT had set up a store account to deal with app submissions from the university, so we were able to use it as a middleman to communicate with Apple on our behalf. That rendered the actual submission process somewhat opaque to us, but it also relieved us of having to deal with the intricacies of packaging, submission, possible rejection and resubmission, and other store-related minutiae.

Reaction

Since being accepted and listed on the iTunes Store in early 2010, the CUL app has been downloaded more than 2,000 times, with a weekly average of between 20 and 30 downloads. Since the new version was released, more than 700 users have created PINs (personal identification numbers) to use the app's authentication features. Feedback has been largely positive, including comments such as "beautiful. period" and "a truly beautiful app . . . actually amazing!!" Blogger Gerry McKiernan, who writes about mobile technology in libraries, noted that the app has "the most functionality of any library app" he knows (personal communication, January 7, 2010).

While we have not carried out any formal assessment of the service, anecdotal feedback and download statistics show that the app has been widely tried and, for the most part, positively received.

AFTER THE LAUNCH

Additional Student Development

For our first student development worker, we hired one of the students who had been on the original development team. It was a natural choice, as he was already an experienced iPhone developer and familiar with the app and its code. His workflow and relationship with the library remained fairly loose, following the pattern established during the original project: An LOL representative and a developer (serving as the student's official supervisor) met with him every few weeks to discuss new development work; then the student would go off and complete the work on his own equipment and schedule.

Over the course of the first academic year in which our student was maintaining development, he handled several issues to improve the app. One item of importance involved fixing bugs. We tried to catch these ourselves, but sometimes they would be reported by users through feedback links on the CUL website or—to our slight dismay—by posting negative reviews on the iTunes Store.

We added few new features to the app during this time, but there was one that greatly affected user interaction and took considerable time to implement. To let a user of the app view his or her personal account (showing requests, items charged, and any accrued fines or fees), the app must authenticate the user with a CUL server. This means that the app must somehow store credentials that identify the user, something that CIT's policies forbade us from doing. In the original app, we had worked around this problem by using an older authentication method specific to Voyager. This method had been superseded by the newer system but was still available as an entry method to the OPAC. Not long after the app was released, though, a library project was launched to remove the old authentication method from both the OPAC and the iPhone app because it was fundamentally insecure: Anyone could log in and view a different user's account with the right name and identification number.

Our student developer replaced the original authentication on the app with a custom PIN system. New users were required to follow a link from the app to the CUL website, log in using the new universitywide authentication system, and create a PIN solely for use with the iPhone app. This PIN could then be remembered by the app as long as necessary so that a user would not have to repeatedly enter his or her credentials to use it. While effective, this system introduced new complexity into the app and confused some users who had trouble creating and using the PINs.

Later in the same year, Apple released the iPhone 4 with its new "Retina Display." This high-resolution display made the lower-resolution images previously included in many apps look shoddy and unappealing, and Apple urged all developers to replace their images with new, high-resolution versions that would look crisp and clear on the retina display. Our student started the process of converting the CUL app's graphics to Retina standards, but he left before completing the task. This work is ongoing.

Library Development

After a year of development work, our student worker graduated and left the university. Instead of replacing him, the DLIT developer with Objective-C experience who had served as the student's supervisor took the time to maintain the app himself. Since the student left, work on the CUL app has taken on a mostly maintenance-based aspect. Based on the necessity to prioritize a multitude of library projects for a limited number of developers, the app would be maintained but not further developed. Besides the occasional bug fix, this maintenance work has included two significant pieces.

The first was necessitated by a change in the way that CIT handled its iTunes Store account. As iPhone app development became more mainstream, CIT found itself dealing with more and more apps from Cornell in the store, including some that had a purchase price. To better accommodate these changing needs, a new store account replaced the old one. Unfortunately, the iTunes Store doesn't provide a way to transfer an app from one owner account to another. From a customer's perspective, what would happen is that the last version of the app would disappear from the store entirely, and a new version would appear under a different name elsewhere in the store and thus potentially go unnoticed. To handle this problem, we released two versions of the app at the same time: one under the new account as the future upgrade path and a special version under the old account. The special version would display a warning message each time that the app was first launched, explaining the change and providing a link to the new version in the store for downloading.

The other issue pertained to the way in which information about library business hours was retrieved and displayed by the app. From the beginning, we had wanted hours information to be dynamically updated in the app so that it would always be current. We were hindered by the fact that the library website stored hours data in a static way: Changes to schedules and exceptions had to be manually entered, and the display was a static HTML page that did not lend itself to easy screen scraping. So the iPhone app initially relied on a specially formatted XML file ("plist") that was stored on a library server and contained a list of all the standard weekly hours for every library in the system. Users were directed to consult the CUL website for up-to-date information and exceptions.

In early 2012, CUL finally implemented a more dynamically stored and updated central source of hours data that was used to display up-to-date hours on the library website. To improve the hours display on the iPhone app without extensively modifying its code, an external script was written that could query the hours service automatically on a daily basis and rewrite the XML source file on the server to supply up-to-date hours to the app.

As soon as Apple announced the iPad, people started inquiring about the possibility of turning the CUL iPhone app into an iPad app. Although not yet frequent enough to merit inclusion in an FAQ, the subject has come up several times in internal discussions. The CUL app is able to run as an iPhone app on an iPad, but the experience is unsatisfying: The app appears as an iPhone-sized view, looking small in the large screen of the iPad, or it is sized to fit the iPad screen at the cost of making the text and images appear highly pixilated.

Thus far, we have not seriously contemplated a project to build an iPad version of our app. A good iPad app is not a simple conversion of an iPhone app but rather a careful consideration of how best to use the increased screen space and different ergonomics. It would take another project of the scale necessary to create the original iPhone app to create a good iPad app.

Although future work on the app is currently limited to maintenance and bug fixes, developments elsewhere in the library may require changes to some of the app's functionality. CUL is beginning a major overhaul of its main catalog and search interface on the library website. This may well mean that the catalog interface in the app will have to change as well to provide full functionality to our users.

CAVEATS AND PITFALLS

Using CIT as a go-between made the iTunes Store submission process relatively painless. It was occasionally frustrating not to have any direct knowledge of what was going on in the process, and we relied on a third party's schedule to submit new versions of the app. However, the benefits outweighed the costs. Not having to deal directly with the iTunes Store relieved us of the burden of having to learn the ins and outs of the application process, and having a contact who was sympathetic to the library's interests and more knowledgeable about iPhone app development and submission than we were was a great benefit at times.

Getting new versions of the app into the store has been a mostly painless process. Our first submission was rejected for using some standard-issue interface icons in nonstandard ways. Once we amended the problem areas, the app was accepted quickly. The only other wrinkle came when we changed the authentication system used by the app: We learned to always supply a set of log-in credentials to Apple along with the new app so that it could see what information lay behind the log-in system.

One of the bigger and more unexpected stumbling blocks we encountered after the development project was finished involved the copyright of the code. The LOL team had not considered the question of who owned the code during the development process, but once we had a finished app, it suddenly became a significant issue. Under Cornell's guidelines, the code's copyright was actually held by the student developers themselves. We had no intention of selling or commercializing the app, but we wanted to have the freedom to share it with others or release the code as open source if we chose to do so. We ended up having to contact all the student developers again and have them sign a document licensing the code to the library to use under fairly broad terms, with the one caveat being that we would not use the code for commercial purposes. Fortunately, they were all willing to sign!

Even during the initial development of the app, a few people questioned why we had chosen to develop an app specific to one platform. One CUL developer wondered when we would write a similar app for the Android system, for example, even though Android was still in a nascent phase at that time. Of course, in the years since the app was first created, use of smartphones has exploded in general, and Android in particular has grown to rival—some would argue exceed—Apple's platform. The iPad has also started a similar growth spurt in the newer tablet market. Thus, it's worth asking whether we would still choose to develop an iPhone-only app if we were starting over again today.

Most likely, we would not. Three years ago, favoring the iPhone was a plausible move given the state of the market at the time. Today, with a wider assortment of options having gained market share, it would be harder to justify their exclusion. Likewise, it would not have been feasible at the time for the library's limited developer resources to be applied to building different apps for multiple platforms. Some developer tools offer the promise of "write once, compile many times" to allow one set of code to run on multiple platforms (e.g., PhoneGap[3]). However, with the growing adoption and greater abilities offered by HTML 5, the best solution today might be to write a web app that would reside on CUL's servers and respond with properly formatted output to any phone accessing it. Concurrent with the iPhone app project, the library created a mobile-friendly version of the CUL website for those not using the app; a future solution might combine the benefits of both approaches.

The CUL's (2012b) purpose statement includes this declaration of its future goals: "As CUL anticipates the future, it is successfully using the latest tools and technologies to make its growing collections more readily accessible to users across campus, and indeed, around the world." The iPhone and its associated applications have certainly made a huge splash on the technology scene. It was a thrill for the library, in cooperation with our student developers, to harness this cutting-edge technology to bring some of its most valuable resources and services to its diverse user population. It seems likely that other academic libraries could adapt our techniques to launch beneficial collaborative projects with their students.

NOTES

1. http://www.youtube.com/watch?v=-ygOBpAY7WA.
2. http://blogs.cornell.edu/theessentials/2011/02/08/isecurity.
3. http://phonegap.com.

REFERENCES

Chapman, S. (2009, January 23). *Two new mobile library interfaces.* Retrieved from http://userslib.com/2009/01/23/2-new-mobile-library-interfaces

Cornell University Library. (2012a). *Goals and objectives.* Retrieved from http://www.library.cornell.edu/aboutus/inside/goals

Cornell University Library. (2012b). *History and purpose.* Retrieved from http://www.library.cornell.edu/aboutus/inside/history

Glazer, G. (2010, March 15). Mobile site and iPhone app make Cornell Library portable. *Cornell Chronicle.*

Kraft, D. (2009, April 27). Mobile devices and CUL. *CULTech Blog.* Retrieved from http://cultech.library.cornell.edu/2009/04/mobile-devices-and-cul.

Merola, R. (2010, February 16). Computer science students to unveil university library iPhone application. *Cornell Daily Sun.*

Read, B. (2009, March 4). Duke U. unveils application suite for iPhone. *Chronicle of Higher Education.* Retrieved from http://chronicle.com/blogs/wiredcampus/duke-u-unveils-application-suite-for-iphone/4557

Sierra, T., Ryan, J., & Wust, M. (2007, December 17). Beyond OPAC 2.0: Library catalog as versatile discovery platform. *Code4Lib Journal.* Retrieved from http://journal.code4lib.org/articles/10

ADDITIONAL READING

Connolly, M., Cosgrave, T., & Krkoska, B. B. (2011). Mobilizing the library's web presence and services: A student–library collaboration to create the library's mobile site and iPhone application. *Reference Librarian, 52*(1/2), 27–35.

Cosgrave, T., Bongort, K., Krkoska, B. B., Silterra, R., & Connolly, M. (2010, February). *Mobilizing the library's web presence and services: A student–library collaboration to create the library's mobile site and iPhone application.* Paper presented at the Handheld Librarian Online Conference II: Alliance Library System and Learning Times.

Kroski, E. (2008). On the move with the mobile Web: Libraries and mobile technologies. *Library Technology Reports, 44*(5), 1–48.

2

Launching a Mobile Initiative: Outreach Strategies

ALEXANDRA W. GOMES
Himmelfarb Health Sciences Library

Anyone who has walked down a city sidewalk, ridden a subway train, or passed through an airport in the last few years has likely noticed the rapid increase in the use of smartphones. They have become an integral part of many peoples' lives, functioning as a communication device, information resource, personal productivity/recreational tool, and entertainment center in a convenient portable format.

In a February 2012 survey by the Pew Research Center's Internet and American Life Project, 46% of all American adults own smartphones. However, within specific age groups, the percentage is even higher. Sixty-seven percent of 18- to 24-year-olds and 71% of 25- to 34-year-olds are smartphone owners (Smith, 2012b). Further surveying uncovered that 75% to 80% of 18- to 34-year-olds who own cell phones use those devices to go online (Smith, 2012a) and 86% of smartphone owners of all ages are utilizing them for research to meet just-in-time information needs, such as solving unexpected problems or settling arguments (Rainie & Fox, 2012).

However, long before the masses began adopting smartphones, the medical field understood the advantages of a portable device that could supplement an individual's recall of medical knowledge and so replace pockets full of note cards. The patrons of Himmelfarb Health Sciences Library at the George Washington University (GW) recognized this functionality at an early stage, and in the early 21st century, early adopters brought their Palm and Treo devices to the library to inquire about compatible medical resources. At that time, electronic books and journals were in their infancy, and few of them were providing content in a format compatible with personal digital assistants (PDAs).

Ten years later, the environment has changed dramatically. Smartphones have replaced PDAs, providing greater functionality, speed, and connectivity. The university's wireless network has been implemented and has grown to include the entire campus. Authentication to university resources is now possible through a virtual private network (VPN) client for laptops, and the Himmelfarb library's collections have transformed from a largely print collection to a predominantly electronic collection. However, just as in earlier days, students, residents, and physicians are still coming to the library asking for assistance in accessing these electronic resources from their mobile devices and acquiring and installing useful medical apps.

The Himmelfarb Health Sciences Library supports faculty, staff, and students in the School of Medicine and Health Sciences, the School of Public Health and Health Services, and the School of Nursing. In addition, medical residents in 37 graduate medical education programs, as well as staff in the affiliated hospital and faculty practice group, are also served by the library's staff and collections. Beginning in 2007, the library began a multiyear project to convert the majority of the library's collection to an electronic format. In 2012, the Himmelfarb library provides more than 3,500 electronic journals, 1,100 electronic books, and 100 databases. A small print journal collection remains, along with print reference and monograph collections; however, the emphasis for both acquisition and use is on electronic materials.

SEIZING THE OPPORTUNITY

In the spring of 2010, library staff recognized the need to develop an organized approach to supporting patrons in the mobile environment at GW. The growth of the library's electronic resources was mirrored by similar growth in the availability of mobile companion products, both apps and mobile-optimized websites, bundled with the library's licenses. However, the catalyst for this initiative was the announcement from the university's information technology department regarding the upcoming release of a mobile VPN client for fall 2010. The combination of numerous mobile devices, the growing mobile medical resources collection, and the impending availability of mobile authentication necessitated a plan to meet the anticipated surge in requests for access and assistance.

Assembling the Team

To begin the planning process, a mobile committee was formed, chaired by the library's webmaster. Committee members were selected on the basis of library roles and expertise, as well as interest in the mobile platform. All areas of the library were represented to include a variety of skills and perspectives on the proposed initiatives. Committee members included the electronic resources librarian, who possessed extensive knowledge of the bundled mobile apps and mobile-optimized websites available through library subscriptions; a reference specialist, who received many requests for assistance on mobile devices at the reference desk; the web services assistant; and a few other staff members, who were users of mobile devices in their private lives and therefore more familiar with typical questions and problems. The size of the team was limited to five or so people to maximize ideas and perspectives as well as allow for easier consensus and communication. The committee's goals were to assess the Himmelfarb mobile environment, recommend a plan to meet the needs of library patrons in this environment, and then implement the plan.

Assessing the Environment

The mobile committee began by conducting a survey of student mobile usage and device preferences. During the summer, the online survey was sent to medical, health sciences, and public health students requesting information about their devices and mobile habits. Although the response rate was low (102 students out of approximately 2,000 students), it provided a snapshot of the student mobile environment. The survey results showed that the most popular devices among students of any school were iPhones and iPod Touches, followed by Android and Blackberry devices. The majority of respondents indicated that they expected to use a mobile device to access medical or public health resources in the future, and nearly two thirds of students said that they would use their device for reference assistance via text, e-mail, or phone. Nearly 90% of survey participants wanted the ability to use their mobile device to renew books or place holds on items in the catalog. Both medical and public health students reported use of mobile websites ranging from Google and *New England Journal of Medicine* to Epocrates and Dynamed.

To supplement the information gathered in the survey, members of the committee reviewed numerous existing mobile medical library webpages as well as other mobile webpages. Reviewers were asked to gather information on content, layout, and design. Mobile websites for peer institutions were also studied, along with government mobile websites, such as MedlinePlus and the Centers for Disease Control and Prevention. Each committee member presented his or her research to the group, along with opinions on good features and useful content items. Meetings took place in a conference room with an Internet connection and projection capabilities so that all members could view specific websites while they were being discussed.

An important part of the environmental assessment was completed by the electronic resources librarian. Each electronic resource held by the Himmelfarb library was evaluated for the existence of a mobile app included in the current Himmelfarb license or a mobile-optimized website. Numerous apps and websites were identified, along with an interesting variety of methods for accessing them (individual serial number, personal log-in/password, etc.). The Himmelfarb library catalog was also examined for its utility in the mobile environment. The Sirsi software implemented in the library was an older version that did not include some of the newer features. This version lacked the mobile functionality that would allow patrons to easily view the catalog and place holds through a mobile interface. Patrons could still access the catalog from the library website, but the text was small and difficult to read when displayed on the small screen.

Creating the Plan

The overarching goal of the mobile committee was to put the library in a proactive position so that when patrons were finally able to connect and authenticate to GW resources through their smartphones, the library would have resources and services in place to support them.

The mobile committee met multiple times to share information gathered from the surveys, mobile website research, and Himmelfarb mobile collections. They were strongly influenced by student survey comments on predicted mobile use of library resources, as well as past library experiences with patrons using earlier PDA devices. The library also wanted to meet anticipated patron expectations for assistance and functionality in this mobile arena.

The resulting plan had several parts:

- develop a central location on the library website to present the mobile resources (apps, mobile-optimized websites) in an organized fashion, separate from the current listings of electronic databases, textbooks, and journals;
- develop a mobile version of the library's website to meet anticipated use from library patrons;
- provide a series of drop-in office hours for one-on-one assistance with mobile devices; and
- advertise the mobile resources and services available to patrons.

IMPLEMENTING THE PLAN

Mobile Resources LibGuide

Since the electronic resources librarian was most familiar with the array of mobile apps and optimized websites included in the Himmelfarb subscriptions, she agreed to tackle their organization. The library had recently acquired a license for LibGuides and started developing individual guides for programs and specific classes containing subsets of library resources (books, e-texts, e-journals, e-databases, websites). The software was easy to use and did not require any knowledge of HTML or web programming. The LibGuide pages could contain text, images, links to e-resources, as well as added features, such as an embedded widget for the reference desk chat program and contact information for the authoring librarian. Since the mobile resources were also a subset of the larger collections, pulling them together in a LibGuide seemed a good fit. Its simple interface also meant that the librarian could integrate the creation of the Mobile Resources LibGuide into her own workflow without needing to coordinate with the webmaster to develop a more formal website.

After some discussion in the mobile committee, the Mobile Resources LibGuide (Figure 2.1) was organized into four parts: Websites and Apps, Calculators, Utilities, and Wireless Access and Device Tips. The guide was envisioned as a place that patrons could visit from their desktop or laptop computer to learn about the resources available and how to add them to their device.

The main focus of the site was the list of apps and mobile-optimized websites that were part of Himmelfarb's electronic subscriptions. These resources were grouped by format (app or optimized website) and then listed alphabetically. A short description was added, as well as resource-specific information on downloading and installing each app. The installation information was included because of the wide variety of steps required. Some resources required users to set up an account on the resources' desktop-formatted website and then access the account from their mobile device. Other resources required the library to distribute a unique serial number to each user, which was then used to install the app. This approach also required the library to develop a system for distributing the serial numbers. Finally, a few free apps were included in the list because of their popularity (Diagnosaurus, Epocrates) and could simply be added by visiting the appropriate app store.

Because the library's patrons reported using a variety of mobile devices (e.g., iPhone, Android, Blackberry), it became important to indicate which apps were compatible with specific platforms. The electronic resources librarian utilized icons next to each app to provide a quick visual clue to the patron looking for suitable apps.

Committee discussions regarding the content of the Mobile Resources LibGuide and the library's mobile website took place concurrently, so several categories of resources were chosen for inclusion on both sites. Several mobile-compatible calculators (free and fee based) were identified and included,

Himmelfarb Health Sciences Library

Research Guides

Himmelfarb Health Sciences Library » Research Guides » Mobile/Handheld Resources

Mobile/Handheld Resources

Last Updated: Nov 6, 2012 | URL: http://libguides.gwumc.edu/mobile | 🖨 Print Guide | 🔊 RSS Updates | ❂ SHARE ⬛ 💬 ✉ ...

| Health Info Websites & Apps | Calculators | Utilities | Wireless, Email & Clinical Systems | HimmelfarbMOBILE |

Health Info Websites & Apps 💬 Comments(0) 🖨 Print Page

Search: [ ] [All Guides ⌄] Search

Mobile Applications / Downloads

- AFP By Topic: Editors' Choice 📱 📓
  Continually updated, selected content from American Family Physician includes pertinent AFP articles and departments, summaries of practice guidelines from major medical organizations, patient education handouts, articles from Family Practice Management, and American Academy of Family Physicians' Metric practice improvement modules. **Access:** Free

- Best Practice 📓
  Evidence-based resource providing disease and drug information. Disease information uses step-by-step approach, covering prevention, diagnosis, treatment and prognosis. **Access:** Courtesy of Himmelfarb Library. **Instructions:** access Best Practice then login or register for a personal account, download BMJ Best Practice from the Apple iTunes Store, login to the app with your personal login/password to download topics and then download images.

- Diagnosaurus 📱 📓 ⣿ ⬤ 📲
  Differential diagnosis tool allows users to navigate by diagnosis, symptom, or organ system. **Access:** Free or minimal fee (.99 for

Mobile-Optimized Websites

- AccessMedicine Mobile ⬤
  Full-text clinical and basic sciences textbook collection that includes Harrison's Online, Tintinalli's Emergency Medicine, Current Medical Diagnosis and Treatment, Hurst's The Heart, Goodman & Gilman's The Pharmacological Basis of Therapeutics, and the Lange educational library series. **Access:** Courtesy of Himmelfarb Library. **Instructions:** access AccessMedicine from desktop computer and create a 'My AccessMedicine login/password which can be used to login to AccessMedicine Mobile on your handheld device.

- Best Practice ⬤
  Evidence-based resource providing disease and drug information. Disease information uses step-by-step approach, covering prevention, diagnosis, treatment and prognosis. Drug information provided by AHFS Drug Information. **Access:** Courtesy of Himmelfarb Library. **Instructions:** Access Best Practice from desktop computer and create a 'My Best Practice login/password which can be used to login to Best Practice Mobile on your handheld device.

Subject Guide

Laura Abate

Contact Info
Himmelfarb Health Sciences Library
The George Washington University
2300 I Street. N.W.
Washington, DC 20037
Phone: 202-994-8570
Fax: 202-994-4343

Figure 2.1. Mobile Resources LibGuide.

as well as a few utility programs that allow the user to read documents on their mobile device.

In late summer, when the university launched the mobile VPN client for authentication, the last page was added, providing instructions on accessing the GWireless network. The result was an easy-to-navigate, organized, single location providing information about the library's mobile resources, including content, access instructions, and platform compatibility. Primary responsibility for its development was undertaken by the electronic resources librarian, with feedback provided by members of the mobile committee and the reference team during its development.

Himmelfarb Health Sciences Library Mobile Website

While the Mobile Resources LibGuide was meant to be a consulted resource, the Himmelfarb mobile website (Figure 2.2) was meant to be a tool, allowing

Figure 2.2. The Himmelfarb Health Sciences Library mobile website.

patrons to complete tasks from their mobile devices. The mobile committee considered developing a mobile app in lieu of a website, but time constraints plus the lack of a university-sanctioned official device encouraged the committee to pursue development of a platform-independent mobile website.

Given the committee's research, knowledge of the Himmelfarb mobile collection, and anticipated patron tasks, the group formulated a few guiding principles for the website. It should have substantial white space and minimal text, focus on expected high-use resources and utilities, link users to the full library website if needed, and provide access to the reference desk for assistance.

Following these principles and incorporating the discussions about content for the LibGuide, the following categories were established for the website:

- Databases/E-Texts
- Calculators
- Drug Resources
- Library Catalog Requests
- Utilities
- Ask Us
- About Himmelfarb

Just as in the LibGuide, the Databases/E-Texts section was divided into apps and mobile-optimized websites. A footer was added to the mobile website containing links to the full library website, as well as the full name, address, e-mail, and phone number for the library.

The library's webmaster took the lead on programming the mobile website, utilizing the categories established by the committee and the links to mobile resources provided by the electronic resources librarian. Small icons were also developed for the website categories to provide visual cues regarding their content, and a library brand was developed for the site (HimmelfarbMOBILE).

Once the initial site was completed, the mobile committee put out a call for student participation in focus groups. The committee hoped to get feedback on the usefulness of both the included resources and the organization of the site. Meeting days and times were carefully chosen to match breaks in student class schedules, and solicitations were tailored to each school or program. Unfortunately, despite several requests, no one responded to the invitations.

The final committee discussion revolved around the URL for the mobile website. At the time of development, only two mobile websites existed elsewhere on campus, and there was no standard naming convention. The committee debated between adding a suffix to the library webpage URL (http://www.gwumc.edu/library/mobile) or inserting a prefix (http://m.www

.gwumc.edu/library). Both approaches were utilized by mobile websites at that time. The group decided that the suffix option would be less confusing and easier for patrons to remember.

Once complete, the website was linked from the library's home page and the Mobile Resources LibGuide.

Office Hours Support

Over the years, Himmelfarb patrons had developed a perception of the library as a wealth of information resources as well as a technology-savvy staff. The reference desk had fielded questions about PDA resources and wireless connectivity for years before the library was able to provide those resources and services. The mobile committee expected that patrons would continue to turn to the library for assistance in these areas, and it wanted to devise a system for handling those requests. Patrons were always welcome to inquire at the reference desk, but with the launch of the mobile VPN client, library staff expected a marked increase in requests for assistance and connectivity.

The mobile committee decided to offer a series of advertised drop-in sessions in which users could receive one-on-one assistance with their mobile device. Library staff could help install apps for library resources, connect to the university's wireless network, install and configure the mobile VPN client, and answer any other mobile- or library-related questions from the user.

Four sessions were initially scheduled for the fall semester. Based on the open office hours model, they were planned to last for 2 hours and were staffed by library staff members who had their personal Apple or Android devices. These staff members, from a variety of library divisions, were selected for their mobile knowledge and familiarity with mobile devices and app installation procedures. The sessions were targeted to medical, public health, and health sciences students and scheduled for times during breaks in their daily class schedules.

The sessions were popular with the students, and attendance ranged from 4 to 10 people. Two additional sessions were scheduled to accommodate the demand for assistance. Students reported that the one-on-one assistance was the most important element of the sessions, since it allowed for an interaction tailored to their specific needs and questions.

Marketing

Both the Mobile Resources LibGuide and the HimmelfarbMOBILE website were launched in September 2010. To make patrons aware of the new mobile resources and services, the library developed a marketing campaign. At the beginning of the fall 2010 semester, e-mail messages were sent to the medical student listservs advertising the lunchtime office hours sessions and the existence of the mobile VPN client and medical apps. Additional e-mail messages were sent to a broader student e-mail listserv promoting the late-afternoon drop-in sessions (3:00–5:00 PM), which were expected to bring in public health students prior to their afternoon and evening classes.

In addition to the e-mail messages, a large poster was placed in the library's lobby on office hours days promoting the service and listing the designated days and times for all the drop-in sessions. Smaller versions of this poster were also posted throughout the library. The library's blog and quarterly newsletter also ran stories about the new Mobile Resources LibGuide and mobile library webpage. These resources were also included in orientations for new students in public health and nurse practitioner programs. Finally, a screensaver was created and added to the screensaver image rotation installed on all public workstations in the library.

ASSESSING THE PLAN AND ADDRESSING THE OUTCOMES

Mobile Resources LibGuide

The mobile committee's efforts in planning the library's mobile approach were highly successful. As a result of these efforts, the library was able to proactively respond to users' expectations for resources as the university expanded wireless access to the mobile arena. The library was well positioned to showcase the staff's knowledge and mobile library resources.

The Mobile Resources LibGuide was quickly discovered by patrons and had 540 page views during the fall 2010 semester. It consistently ranks as one of the 10 most-visited LibGuides each year out of a current group of 89 LibGuides, and in less than 2 years, it has received a total of 3,654 page views (September 2010–July 2012). The most-visited tab is Websites and Apps (2,454 page views), followed by Wireless Access (689 page views).

The LibGuide continues to change as new resources are added to the collection and additional mobile apps or mobile-optimized websites become available. Later additions to the Wireless tab have included instructions on configuring a smartphone for the GW e-mail system, as well as information on the hospital's wireless network.

After the Mobile Resources LibGuide launch, library staff realized that users were consulting it from their mobile devices and that the table used to display the app information (platform, installation instructions, content description) was too wide to fully display on such small screens. The table was reworked to include smaller icons and a different layout, which now displays appropriately on a mobile device.

HimmelfarbMOBILE Website

The mobile website has also been popular, with 291 page views in fall 2010 and a steady stream of traffic since then. In 2011, the university launched a mobile app for the university and included the library's mobile website in the library section on the home page, adding to its visibility and perhaps accounting for some of the increased traffic.

There have been only minor changes to the mobile website since its launch. A few additional resources were added in the Databases/E-Texts and Drug Resources categories as new resources became available.

Office Hours Support

The popularity of the six fall-semester office hours sessions led the library to schedule six sessions for the spring semester. Session times were scheduled for both noontime and late afternoon to accommodate various student schedules. While student demand for assistance with mobile app installation and connectivity continued, more and more of these requests came to the reference desk or directly to librarians rather than at the drop-in sessions. The last few sessions saw no visits from patrons, and plans for future drop-in sessions were cancelled. The reference desk also experienced a marked increase in requests for mobile assistance after the winter holidays and prior to the start of third-year clerkships, which was attributed to the purchase of new devices as holiday gifts or as part of clerkship preparations.

With the change in mobile request patterns, it quickly became apparent that the support model needed to change. The constant stream of requests for mobile assistance was having an impact on the workflow and workload of the individuals providing mobile assistance to patrons. It was no longer enough to have a few staff members knowledgeable about mobile resources and installation processes; now everyone who staffed the reference desk needed to be trained.

Faced with this situation, the library chose to again take a proactive approach. While some of the library staff members were conversant in mobile technologies, devices, and apps, other members had little or no experience. The decision was made to train all of the librarians, both those in reference and those working behind the scenes. The goals for this project were (1) to increase the librarians' knowledge of mobile apps and installation procedures, thereby adding to the number of individuals who could assist patrons, and (2) to model the use of these mobile resources during the librarians' weekly participation in the first-year medical school's problem-based learning curriculum. If the mobile committee had foreseen this development, another goal for the original mobile plan could have been to train all librarians on mobile devices so that they can model use of the resources and provide support to patrons.

The library was able to purchase 12 iPod Touches and 12 copies of Epocrates Essentials, courtesy of a local grant. The iPod Touch device was chosen because it is a popular operating system but does not require a monthly data plan subscription. Having the same device also allowed for uniformity in training.

Each librarian was issued an iPod Touch and required to attend two training sessions in the summer of 2011. Since the librarians' mobile experience ranged widely, the first session in July focused on basic navigation and app download and installation (DynaMed and Epocrates). All participants were required to set up an iTunes account prior to the training session. Although the session was planned to last only 1 hour, the diversity of mobile abilities resulted in slower progress than anticipated, and even with a 30-minute extension, the instructor was unable to complete the lesson plan.

By the time of the second session in August 2011, the librarians were a bit more comfortable and adept in using their iPod Touches. The lesson plan for the second session was based on input from the librarians and included sharing tips about apps and the devices, registering them with MobileMe in case of loss, and exchanging information about common patron requests and complaints. The instructor received positive feedback about both sessions, but the most significant result is the increased number of staff members who are comfortable providing mobile support to patrons and the decrease in referrals to the original staff members staffing the drop-in sessions.

Marketing

After the initial marketing effort, information about the library's mobile resources and assistance spread rapidly by word of mouth. Students shared knowledge about specific apps with their peers, and there was a constant stream of requests for serial numbers to install popular medical apps.

The library continues to send optimally timed e-mail messages to students, reminding them of our resources (both mobile and otherwise) to support their studies. First- and second-year medical students receive messages at the start of the school year; third- and fourth-year students receive them at the beginning of their clerkship year. Reminders about mobile resources are also sent to all students in January and to second-year medical students near the end of the year, to address the historical rise in numbers of mobile devices at these times.

RECOGNIZING NEW OPPORTUNITIES

There is a constant and rapid rate of change with technology. Platforms change, bandwidth increases, and electronic resources (apps, e-books, e-journals, integrated library systems) change to take advantage of new functionality options. While many factors affect how and when a library incorporates technological change, it is important to be aware of the environmental changes as well as understand the implications for the library. Patrons will continue to span the technology gamut, from early adopters to so-called laggards, but by the time that a particular technology has permeated the late majority adopters, a library needs to have addressed the change somehow.

At the time of this writing, there are several technological changes on the library's horizon. Cloud computing is becoming a factor in integrated library systems. The next version of our library's integrated library system may incorporate the cloud to some degree and provide an easy interface for standard library functions (item renewals, holds, etc.). The iPad and competing tablets are becoming increasingly popular among consumers and library patrons (18% of American adults own a tablet computer; Pew Internet & American Life Project, 2012). While the current mobile website provides a clean interface to library resources and services on a tablet, it does not take advantage of the larger screen size or potential added functionality/integration with other resources or systems. In addition to addressing this new platform, the mobile website is due for a complete review of layout and resources so that it continues to be a current, appealing, and maximally useful website for users. The library's main website also needs to be refreshed to better display the full set of library collections and services whether the user is on a laptop or a tablet. Usability testing through focus groups or individuals will again be a part of both website renovations. Finally, for users accessing the library website from a smartphone, one goal of the website renovation will be to incorporate code that will identify the user's device type and automatically send them to the mobile version of the website if they are using a smartphone or iPod Touch.

CONSIDERATIONS FOR OTHER LIBRARIES

Each library is different and composed of its own collections, staff expertise and knowledge, and mission. All sorts of factors can affect a particular project, including budget, time required, timing in the calendar or academic year, and availability of staff. The original mobile project described earlier in this chapter was completed in a short time by recognizing an emerging need and then marshalling the library's internal resources to meet it. In reviewing the project 2 years later, the importance of a few key decisions is clear.

Take advantage of your staff's strengths and diversity. The mobile committee members' wide-ranging backgrounds proved to be an asset. Everyone was able to contribute vital knowledge about library resources to the discussion and subsequent project regardless of status (librarian, paraprofessional) and library expertise (reference, web services, electronic resources). This diversity also allowed the workload to be divided among the members according to each person's strengths. The librarians were also able to learn from one another's experiences as they became familiar with their iPod Touches in the two instructional sessions.

Know your environment. Because the library had information, anecdotally from the reference desk and statistically from the survey, the committee was able to make informed decisions leading to platform-independent development and the drop-in office hours support sessions. Also, good relationships with the university information technology department resulted in early intelligence about the upcoming mobile VPN client and led to the start of the library's project.

Bring in stakeholders. It is crucial to have the support of those affected by the proposal. Providing stakeholders with updates and the opportunity to convey their perspective allows for buy-in and for additional information that may influence the planning process. The mobile committee kept library administration informed of the proposed plan for the library's mobile presence and, once approved, the progress of the plan's various pieces. Because the mobile committee contained members from all divisions in the library, key personnel were already involved in the planning process. Student input was also sought to improve the content and organization of the developing resources.

Respond to changes. The decline in attendance at the drop-in sessions was unforeseen. Rather than continuing to devote staff time to a service that was no longer useful, a new plan was developed to meet the evolving patron assistance situation. An added benefit of the new assistance model was the opportunity to increase the knowledge base of all the librarians and model use of the mobile resources to the student community.

The library's development of the mobile website, the Mobile Resources LibGuide, and the patron assistance program all serve to promote the library and its resources (collections and staff members). Anticipating and planning for a situation not only fulfills the library's service and support mission but positions the library as a forward-thinking proactive organization, ready to meet environmental challenges as they arise.

REFERENCES

Pew Internet & American Life Project. (2012, August). *Trend data (adults)*. Retrieved from http://www.pewinternet.org/Static-Pages/Trend-Data-(Adults)/Device-Ownership.aspx

Rainie, L., & Fox, S. (2012, May 7). *Just-in-time information through mobile connections*. Retrieved from http://www.pewinternet.org/Reports/2012/Just-in-time.aspx

Smith, A. (2012a, June 26). *Cell Internet use 2012*. Retrieved from http://www.pewinternet.org/Reports/2012/Cell-Internet-Use-2012.aspx.

Smith, A. (2012b, March 1). *Smartphone update 2012*. Retrieved from http://www.pewinternet.org/Reports/2012/Smartphone-Update-2012.aspx

3

Oregon State University Libraries Go Mobile

LAURIE BRIDGES, HANNAH GASCHO REMPEL, AND EVVIVA WEINRAUB

Oregon State University Libraries

In early 2009, about a dozen academic libraries in the United States had websites optimized for mobile devices; however, at that same time 85% to 88% of Americans owned mobile devices (International Telecommunications Union, 2009; M-Libraries, n.d.). Since that time, academic library mobile sites have become nearly ubiquitous. When the Oregon State University Libraries (OSUL) mobile team began looking for examples of best practices in mobile development for libraries in early 2009, the team was grateful for the sites developed by trailblazers such as Ball State University and North Carolina State University. When the OSUL mobile site launched in March 2009, it was unique among library sites because it was optimized for both smartphones and feature phones. Due to the overwhelming interest from other public and academic libraries about our site, members of the OSUL mobile team published several articles about the creation of the site and presented at numerous conferences. While we do not have any magical solutions for developing a mobile site, we have accumulated some expertise and are always planning for the future. In this chapter, we detail the beginnings of our site, its chief aims and objectives, the resources used to develop it, our assessment and usability testing plan, advice for other libraries, and our future plans.

DEVELOPING A MOBILE SITE

In 2004, the OSUL launched an intensive effort to remake the libraries' strategic plan, which included reimagining the mission, vision, and goals of the libraries. One of the items highlighted in the strategic plan was to "pioneer information management tools to enable targeted and rapid information retrieval" (*Oregon State University Libraries Strategic Plan 2004*, 2004). Forging pioneering efforts meant that the OSUL would need to reconsider their staffing mix. It quickly became apparent that without a programmer on board, it would be impossible for the OSUL to realize this new vision. In 2006, a computer programmer/analyst was hired at the request of the librarians. Instead of filling a vacant librarian position, the librarians came together and asked the administration to fill the position with a full-time programmer who would be tasked with addressing the needs and ideas of the reference, instruction, and outreach librarians. The programmer was charged with creating web software in partnership with OSUL librarians, which would make it easier for users to connect with library resources. After little over a year on the job, the programmer teamed up with two librarians to create a mobile website to "deliver information wherever and whenever it is needed."

Before beginning work on the site, the newly formed team was asked to write a proposal for library administration to demonstrate the rationale for developing a mobile site. The proposal included an executive summary, a literature review, a problem statement, and a problem solution, including a project plan with resources, milestones and timelines, screenshots of the full website on mobile devices, and examples of existing mobile sites. The problem statement read, "Patrons are growing accustomed to using mobile phones to access online websites for information, but [OSUL] web pages are too long and cumbersome for small screen viewing." The site was to be implemented over three phases:

Phase 1: Create a mobile website replicating the static pages of the OSUL website.
Phase 2: Develop a mobile application for searching the library catalog and other holdings.
Phase 3: Develop an innovative web application to push out some part of the OSUL collections to users.

Once the proposal was approved by library administration, the team began to develop the new site. The first phase took approximately 3 months and involved creating mobile-friendly pages for the basics: library hours, contact information, frequently asked questions, and directions (Bridges, Rempel, & Griggs, 2010). The second stage of development included somewhat more interactive elements: the library catalog, staff directory, and a computer availability map (Griggs, Bridges, & Rempel, 2009). During the third stage, BeaverTracks (http://tour.library.oregonstate.edu) was created and launched. BeaverTracks (inspired by North Carolina State University's WolfWalk) connects the past to the present by geotagging historic images from the Oregon State University Archives to 22 campus locations to create a mobile walking tour (Griggs, 2011).

There were three guiding questions for every phase of the OSUL mobile site development (Griggs et al., 2009):

Question 1: What are the circumstances of the mobile user?

Question 2: What are the goals of the mobile user?

Question 3: What tasks is the user likely and unlikely to do on a mobile device?

CHIEF AIMS AND OBJECTIVES

One of the original aims was to create a device-agnostic site that could be optimized on both smartphones and feature phones. Because the school is a land grant public university with a sizable first-generation population and international student population, in 2009 the mobile team felt it important to create a site that could be optimized on lower-end feature phones, as well as the popular and more expensive iPhone. A 2010 review of our site statistics revealed that 4% of our users were accessing the site on feature phones, 9% were using DoCoMo phones (a Japanese service provider), 75% were on iPhones, and the remaining 12% were on smartphones other than an iPhone (Bridges & Griggs, 2010). The DoCoMo percentage was surprising to us; we assumed that these users were international students on our campus and that many were on feature phones at the time.

In 2010 with 4% of our users accessing the site on feature phones and another 9% using DoCoMo, the team made the decision to continue optimizing the site for both smartphones and feature phones. However, recent opinion about the continued support of mobile versions of websites for feature phones has shifted as usability experts have shared their mobile strategies. Jakob Nielsen (2011) stopped testing for feature phones in late 2011 and specifically stated on his popular blog *Alertbox*, "Feature phone usability is so miserable when accessing the Web that we recommend that most companies don't bother supporting feature phones." In his book *Mobile First,* Luke Wroblewski (2011) addresses feature phone issues in chapter 1, ultimately concluding that "the mobile industry is moving toward smartphones, and so will this book" (p. 13). Although our mobile team has prioritized supporting feature phones in addition to smartphones, in the next redesign of the mobile site, we may decide to abandon this goal and instead focus where the market is going—namely, to smartphone-only design.

STAFF AND RESOURCES

What does it take to actually build, maintain, and run a mobile website when using your own programmer and information technology staff? When the OSUL initially launched the mobile website, the programmer spent 90% of her time over a 3-month period building the mobile framework. This required her to work with the systems administrator to build the library's first Rack site, a type of middleware that sits between the web application and web server, and to install a lightweight version of Ruby on Rails, the open-source web application for doing the site programming. The systems administrator devoted approximately 30% of his time during the 3-month period to getting the Ruby on Rails applications working and performing tests on the system. Student graphic designers spent over 100 hours modifying and creating logos to work well in a mobile environment. Today, most universities have standardized icons as part of their mobile web-branding program, but at the time that the OSUL initially launched the site, the university was not yet working in the mobile arena; our site was ahead of the curve.

After the initial framework was completed, each piece of the mobile website took less time to setup. During peak project activity times, the programmer would spend approximately 70% of her time over a 1- to 3-month period on various mobile projects. She employed an agile programming model and would outsource pieces of the project to student employee programmers as necessary. Her student programmers, during peaks in the project-building phase, had approximately 20 hours per week of work to do.

Many library vendors offer mobile solutions to access library basics, such as the catalog or chat/text-a-librarian services. In each case, the decision to purchase a solution is something that should be considered in light of the institution's needs. For example, the cost of purchasing the mobile catalog application through the OSUL integrated library system provider was prohibitive; in addition, we were not satisfied with the customization options available via the vendor solution. So instead, the programmer employed a method known as screen scraping to create the mobile catalog functions we desired. While it was not perfect, we were comfortable with the quality of the results pulled from the catalog using this method. Creating an in-house catalog solution also enabled the mobile team to add in special features to the catalog not present in the full-site version, such as immediate identification of the floor location of items and the option to text item records to a phone.

Another example of a product that we chose to develop ourselves was the computer availability map. The standard map is an in-house Oregon State University software program that was not developed by the library, but our programmer modified it to work well in a mobile environment. However, the OSUL did decide to purchase vended solutions for the text and chat programs ("Text a Librarian" and "Library H3lp"), neither of which required programming time for implementation. In this case, the benefit to purchasing a vended solution far outweighed the costs of developing and maintaining a homegrown solution.

The costs of building the site—including the time of the programmer, systems administrators, and students—cost approximately $10,000. The two librarians on the mobile team also worked on this project intensively for a few months by guiding the project and presenting and writing on the process and its outcomes. Their costs are slightly harder to quantify but were likely in the vicinity of $10,000 in total costs. The OSUL did not need to purchase any additional hardware, and all of the software was open source. The total project cost approximately $20,000 to build.

Over the next 2 years, the programmer and her students performed regular testing and updates to the mobile site, making sure that it worked with the largest number of possible devices on the market. The mobile team visited the local AT&T store on two occasions to test the mobile site on all the devices available in the store. The systems administrator occasionally needed to update RubyGems, a software package containing a packaged Ruby application or library, and needed to perform regular systems maintenance. Overall, maintaining this project required approximately 15% of the programmer's time over the 2-year period, approximately 200 hours of the student programmers' time, and less than 3% of the systems administrator's time. The overall cost was in the vicinity of $12,000 for 2 years of support.

In fall 2010, the mobile team decided to redesign the mobile website. This required some work from our graphic design students, primarily in creating icons and revising the color scheme. The redesign also required additional work from our programmer, her student programmers, and our systems administrator, as well as the two librarians. From start to finish, the redesign took approximately 1 month with our programmer working on the project for approximately 70% of her time during that period. Her student employees averaged about 10 hours per week of work on the project, and our systems administrator did not increase his time significantly. With the librarians' time factored in, primarily with site testing, the redesign cost us approximately $3,500.

The biggest costs in creating a mobile website exist in the initial building process. The actual maintenance can be relatively inexpensive when using existing staff, as long as you stay on top of new developments in the mobile landscape and routinely check to make sure that the current site works properly.

NOTES FROM THE FIELD: OUR USERS' REACTIONS

As we described earlier, the development of a mobile site dovetailed well with our library's mission to serve our users whenever and wherever they might be. The OSUL mobile site has received overwhelmingly positive feedback from a variety of communities, inside and outside our university. The library web-developer community appreciated our open-source ethic and willingness to share code and ideas. The local user community also responded to our innovative work. The site was featured in the city newspaper and in a front-page story in the student newspaper. And the dean of student life tweeted, "The mobile library website rocks!" However, a more interesting and helpful way of gauging users' responses comes in the form of personal stories. Here are several anecdotes that illustrate this feedback.

The first is from a student who was using the mobile library site on his iPhone during a library instruction session. One of the authors happened to be leading the library session that day, and the library's teaching classroom was full, so there were not enough computers for everyone to use. While the students were working on the assignment, the librarian circulated throughout the room and stopped to ask this student if he was able to do his work without a desktop computer. He replied enthusiastically that he actually preferred to use the library's resources via the mobile site. He then proceeded to demonstrate how to use the different features on the library's mobile website, unaware that this librarian had been involved with the site's development.

The second anecdote comes from a College of Business faculty member. In her role as liaison, another author was talking with this faculty member, who was unaware that she had worked on the development of the mobile site. He told her, "Hey! I was just about to send you a message. I didn't know you guys had a mobile site! It's great and so easy to use. I wanted to find your phone number and thought it would be difficult, but I just found it so easily. It works really well."

The third anecdote is from a librarian colleague within the OSUL. A department she served was consolidating with another department, which wanted to see how much of its departmental library collection would be useful to our library. Instead of asking the department to create a list of all the books or ship the collection to our library to sort through as she usually would have done, she took her iPod Touch, went to the departmental library, and was able to efficiently sort through the books with the help of the mobile catalog. She said that this experience completely changed the way she would approach future collections tasks and that she now felt untethered from the library in a very positive way.

The last story comes from some consultations that the mobile library team did with the university's central web services department. As mentioned earlier, the library was the first department on campus to release a mobile site. Programmers around campus were impressed with the clean design style and our thoroughness in testing the site. As a result, 9 months after the site was launched, the team working to develop the university's mobile site contacted the libraries' team to discuss the styles, the requirements documents, the mobile site testing plan, and the philosophy behind some of the accessibility decisions made in planning the mobile site. The two teams then worked together to easily connect the library's mobile site to the university's mobile site.

ASSESSMENT OF THE SITE

Earlier in this chapter, we mentioned that the mobile team visited the local AT&T store on two occasions, first during the initial design phase and again during the redesign to test the site on a wide variety of mobile devices in the store. This was just one component of the overall assessment of the mobile site. Early in the development of the site and through each iteration, the programmer used web tools for testing the mobile site, such as mobileOK Basic Tests, MobiReady, browser simulators, and device emulators (Griggs et al., 2009). Other forms of assessment have included routinely checking the usage statistics of the site, which have hovered around 100 daily users for the past 2 years. Usage stats are gathered with Urchin web analytics software, which help us track the pages that are the most visited, the bounce rate of different pages, and the click-through rate. The programmer also did basic catalog search testing by soliciting volunteers from the OSUL staff and then sending out testing directions for volunteers. For example, the Basic Search Test contained the following protocol, to which the volunteer was to provide a written response regarding what happened during that step; volunteers also included their type of device and web browser:

1. Do a quick search by entering a search term and using the default options.
2. Search on the same term a few times changing the branch or number of results options. (If you set the number of results to 1, you should be taken directly to the record rather than a results page.)
3. Pick a different search type, and search for a term using the default options for branch and results.
4. Perform this search again varying the branch and number of result options; compare the results to the initial search.

5. Repeat steps 3–4 for the other search types.
6. Try searching for obscure terms that wouldn't get results (e.g., search for a word using ISN search, search for numbers on a course reserve [rather than the instructor]). You should receive a "no results" message if nothing was found.
7. Try submitting the search form without any search terms; you should receive an error message asking you to input search terms.

ADVICE FOR BUILDING OR EXPANDING YOUR MOBILE SERVICES

If you are at the beginning stages of creating a mobile library site, it is important to first consider some of the unique constraints and opportunities of the mobile context, who your audience is within this context, and what these users prefer to find and use within the mobile context. While there has been an increasing push to move more and more content onto mobile sites, usability experts continue to say that it is best to be selective in what content you choose for the mobile environment (Nielsen, 2012). This may be a particularly important point for librarians who like to provide as much information as possible just in case someone, somewhere, might need it. However, when users interact with a mobile site, they are typically seeking information that is either time saving or location sensitive; they are using tools that expose the native capabilities of the mobile interface, that provide a distraction when they have some time to kill, or that allow them to check in or demonstrate where they are.

In addition, it is important, as with any web-based project, to make sure that you are providing the best user experience possible. To do this, make a point of exploring the different technologies on the market and asking a few key questions:

> *What are your goals with a mobile site?* Are your goals clearly defined and measureable, or are you building a mobile site simply to keep up with other libraries?
> *How will you assess whether your mobile site is successful?* Is number of hits good enough? What do you want to know about your mobile users?
> *What problem will building a mobile site solve?* Are you increasingly observing your users interacting with the Internet via mobile devices, and, as a result, you want to be where they are, when they need you?

In addition to considering the mobile context and the user experience when building or expanding your mobile site, consider your library's strengths. As illustrated in our BeaverTracks example, one of your strengths is likely to be your unique collections. As you think about what to include and feature on your mobile library website, focus on showcasing these unique collections. There is increasing interest in digital humanities, a field where new media and technologies intersect with humanities-based research and teaching. As you think about expanding or augmenting your mobile library's website, consider what collections you have that are already in digital form and how they could be readily incorporated in creative and playful ways by students or other researchers making use of the photography and web-publishing aspects of their mobile devices. What connecting points could you provide to answer new questions or tell stories in a different way?

One of the final strengths that we encourage you to capitalize on is your student population. If you are at an academic library, it is likely that you have skilled students who are able to do some of the heavy lifting in creating the front and back ends of your site. In our case, we were able to use a computer science undergraduate to do some of the behind-the-scenes coding for both the mobile library site and the catalog site. This student worked particularly hard on the catalog site (http://m.library.oregonstate.edu/search/) and in the process learned a great deal about the way that libraries are organized and how information works—a win for both him and the library.

Graphic design student workers were also involved in crafting the icons used on the mobile library site and the overall site mock-ups. One graphic design student was heavily involved in the process of designing our Book Genie mobile book recommender site (http://genie.library.oregonstate.edu). This process gave him the experience of working with a client and turning someone else's vague ideas into a clear end product. Using students is a smart decision when working on a mobile library site, not just for the specific technical skills they can bring, but also the user perspective they can provide. These students are likely to use the mobile web differently than librarians and, as a result, can bring those perspectives to the design and development process. In addition, they use the library website for their own coursework and know what will actually be helpful for them as students.

NEXT STEPS FOR THE OSUL MOBILE SITE

The OSUL mobile team has had several personnel transitions over the past couple of years, and in the winter of 2012, the team decided to move the responsibility for maintaining our mobile website from the programmer to the web services coordinator. This transition period has given us the opportunity to reexamine what shape we would like our site to take next and to make sure that we are still on top of the steady stream of changes to mobile technology. When the site was launched in 2009, HTML5 was still being talked about in whispers, and Drupal wasn't ready for prime time. Now, with both of those platforms in full production, we will explore the possibility of using either HTML5 or Drupal as our mobile web platform, and we will look at the potential of moving away from Ruby on Rails. Moving to Drupal would allow us to couple the content with our existing web management software—saving time and money—whereas moving to HTML 5 would allow us to provide "offline mode" and play videos. However, there are still tremendous benefits to remaining with Ruby on Rails for the mobile environment, not the least of which is leveraging the RESTful technologies it employs. Each benefit and pitfall of the alternatives need to be weighed, as making the decision to migrate to a better tool might be costly and time-consuming in the short run, but the shift might offer greater staffing flexibility or better interoperability in the long run.

Another area that is rapidly expanding for both mobile web development and libraries is digital publishing. We have begun to focus more of our attention in that arena. In April 2012, we launched the first of many planned collaborations with Oregon State University Press—a mobile walking tour called "Bart King's Architectural Guidebook to Portland," a companion to the print edition by the same name. This tool provides images, information, and geolocation services and can be found at http://pdxarchitecture.library.oregonstate.edu.

We have also begun exploring the creation of a mobile framework that will allow us to easily mount digital collections in a mobile environment. Our plan is to release the framework for that project on a number of code-sharing sites, along with documentation on how to pull data from an institutional repository through the framework relatively easily.

Finally, we are working on a redesign of our mobile website separate from the platform that we use for delivery. As part of the mobile movement that is trying to provide access to content as quickly as possible, we plan to move a search box to the mobile home page (Wroblewski, 2011). We will begin the process by developing several design options, and then we will conduct usability testing using hybrid field and laboratory methodology. Throughout all of these changes and testing, we will continuously seek feedback from our users, and we will monitor the new opportunities that the mobile context sends our way to provide the best possible mobile library experience. We would also like to thank Kim Griggs, our programmer from 2006 to 2011.

REFERENCES

Bridges, L., & Griggs, K. (2010, July 23). *Handheld librarian III*. Retrieved from http://www.slideshare.net/ilaurie/handheld-librarian-iii

Bridges, L., Rempel, H. G., & Griggs, K. (2010). Making the case for a fully mobile library web site: From floor maps to the catalog. *Reference Services Review*, *38*(2). Retrieved from http://hdl.handle.net/1957/16437

Griggs, K. (2011). Geotagging digital collections: Beaver Tracks Mobile Project. *Computers in Libraries*, *31*(2), 16–20.

Griggs, K., Bridges, L. M., & Rempel, H. G. (2009). Library/mobile: Tips on designing and developing mobile web sites. *Code4Lib Journal*, *8*. Retrieved from http://journal.code4lib.org/articles/2055

International Telecommunications Union. (2009). *Information society statistical profiles 2009: Americas*. Retrieved from http://www.itu.int/pub/D-IND-RPM.AM-2009

M-Libraries. (n.d.). *Library success: A best practices wiki*. Retrieved from http://www.libsuccess.org/index.php?title=M-Libraries

Nielsen, J. (2011, September 26). Mobile usability update. *Alertbox*. Retrieved from http://www.useit.com/alertbox/mobile-usability.html

Nielsen, J. (2012, April 10). *Mobile site vs. full site (Jakob Nielsen's Alertbox)*. Retrieved from http://www.useit.com/alertbox/mobile-vs-full-sites.html

Oregon State University Libraries strategic plan 2004. (2004). Retrieved from http://hdl.handle.net/1957/7991

Wroblewski, L. (2011). *Mobile first*. New York: A Book Apart.

4

Making the Library Mobile on a Shoestring Budget

Helen Bischoff, Michele Ruth, and Ben Rawlins
Georgetown College Library

WHY MOBILE MATTERS

As handheld technology becomes more prevalent and affordable, libraries—whether classified as public, academic, or special—will likely find that it is strikingly common for adult library patrons to own handheld devices, such as mobile smartphones and tablet devices. According to a study conducted by the Pew Internet and American Life Project, "88% of U.S. adults [owned] a cell phone of some kind as of April 2012, and more than half of these cell owners (55%) use their phone to go online" (Smith, 2012). What's more important to note, as far as libraries are concerned, is that 31% of the cell Internet users use their cell phones almost exclusively for going online. Cell phone Internet usage can certainly affect the way in which library patrons choose to access and use online library resources.

In addition to cell phones and smartphones, the adult ownership of tablet and e-reader devices is also on the rise. From December 2011 to January 2012, adult ownership of either an e-reader or tablet computer increased to 19% from 10%, and "the number of Americans owning at least one of these digital reading devices jumped from 18% in December to 29% in January" (Raine, 2012). These mobile statistics matter to libraries and mattered greatly to the Ensor Learning Resource Center (LRC) at Georgetown College,[1] because handheld devices have greatly affected current library services and will undoubtedly change the way in which libraries continue to provide services to patrons in the future.

Specifically, the fall 2011 Georgetown College information technology services department statistics showed that of the 2,100 campus members, the campus network registers 870 mobile devices (this includes smartphones,

**Table 4.1.   Number of Registered Mobile
Devices on the Campus Network**

Operating System	No.
Apple iPhone	486
Apple iPod	198
Apple iPad	124
Android	62

Note: Data compiled from the fall 2011 Georgetown
College information technology services survey.

tablet computers, and e-reader devices). Table 4.1 shows the breakdown of the devices by mobile operating system and the number of registered clients on the wi-fi network at Georgetown College. Notably, these statistics do not reflect the campus members who may own mobile devices but use their providers' data plans exclusively. Thus, there are likely many handheld devices, especially smartphones, that aren't registered to the campus network but are still accessing the Internet, online library resources, and other online campus resources on a daily basis.

Given the increase in cell phone Internet usage and the spike in ownership of e-readers and tablet devices, the Ensor LRC has made a conscientious effort to provide services and resources that cater to this trend. The following sections outline new programs and enhanced services, along with information on how to set up said services, that the Ensor LRC has established in response to the growing demand to make more resources mobile friendly and accessible.

MARKETING EXISTING RESOURCES

The increase in cell phone ownership and cell Internet usage, along with tablet and e-reader ownership, has affected the way in which the Ensor LRC provides reference and instruction services to its college community. The traditional face-to-face delivery of reference assistance and library research instruction is no longer as relevant to Georgetown College students as was perhaps the case 10 to 15 years ago—thus, the need to change. Pending budget considerations and the availability of open-source software and programs, the Ensor LRC reference and instruction services are slowly evolving with these technological advances and changes.

A breakdown of instruction and reference/research transactions at the Ensor LRC was telling as far as how students and faculty were seeking assistance from the library (Table 4.2). In the 2010–2011 school year, Ensor LRC

Table 4.2. Number of Library Instruction Sessions and Attendees

School Year	No. of Instruction Sessions	No. of Attendees
2010–2011	55	1,012
2011–2012	58	822

Note: Data compiled from the Ensor Learning Resource Center's library instruction statistics, 2010–2012.

taught 55 face-to-face and online webinar research instruction sessions with 1,012 attendees (mostly undergraduate students with some graduate students and campus faculty/staff). In 2011–2012, the number of instruction sessions increased, but the number of attendees decreased to 822 attendees. This was due in large part to a decrease in enrollment and general education curriculum changes, and with the emergence of more online courses, fewer faculty members were requesting research instruction. It seemed that faculty overestimated the research skills of online students—that these students must inherently be better researchers (when it came to the library's online resources) by virtue of taking online courses.

The Ensor LRC Research Help Desk (traditionally known as a reference desk) also experienced similar decreases (Table 4.3), with 1,699 transactions in 2010–2011 down to 1,132 transactions in 2011–2012. Notably, the percentage of technology-related transactions showed a slight percentage increase, from 69% in 2010–2011 to 72% in 2011–2012.

The increase in questions dealing with computers, tablets, e-readers, smartphones, software, equipment, and so on, seemed to indicate to our research team that while students certainly held the latest gadgets and technology at their fingertips, they still needed guidance in how to use these resources in a research and educational context. Therefore, the Ensor LRC set out to integrate and promote its library resources to coincide with students' and faculty members' mobile devices.

Another statistic that gave rise to the need to connect with students on a broader scale and in greater mobile and virtual contexts was the increase in our online graduate student enrollment. From fall 2002 to fall 2011, the graduate education programs have increased enrollment from 342 students to 545 students—a 63% increase (Georgetown College, 2011). While most students lived within the state of Kentucky, the majority did not live within

Table 4.3. Number of Research Help Desk Transactions

School Year	Total No. of Transactions	No. of Technology-Related Transactions
2010–2011	1,699	1,178 (69%)
2011–2012	1,132	814 (72%)

Note: Data compiled from the Ensor Learning Resource Center's library instruction statistics, 2010–2012.

a 30- to 50-mile radius of the physical campus. Early in the program's history, students met on campus in a traditional face-to-face format with some classes online, whereas now all graduate education programs are available online. The online evolution of the graduate education program also necessitated the need for the library to keep pace with and promote virtual and technical library resources.

In response to the increased number of online students, the research team decided to purchase a LibGuides subscription in fall 2009 from Springshare Company. In the past, research guides had been created and distributed via paper handouts in library instruction sessions, or there was a general research page on the library website. Though these older methods of informing students about research strategies could still work, we were having a difficult time reaching a broader student audience, and many of the handouts made their way to a trash can either by students' doing or because information became outdated along with database interface screenshots. Also, the creation and publication of such research aids to the library website had to go through a single webmaster, which often delayed the process of posting research tutorials and made creating interactive research resources more cumbersome and time-consuming. Now, the research team uses LibGuides as part of regular library instruction, so every professor/course that requests research instruction or support gains a dedicated research page. Our library liaison program, wherein each librarian is assigned to work with a set of academic departments, also uses LibGuides to create research modules for the different academic disciplines. The ease of creating an online research module changed, literally, overnight with LibGuides and has enabled the research team a far greater reach within the Georgetown College population, but it has also opened up opportunities for us to create research guides in partnership with outside libraries and other related programs.

In addition to LibGuides, the research team implemented and integrated an online chat service into the library's webpage and LibGuides subscription in January 2010. At the time, Meebo Chat was free and therefore an obvious choice for testing online research support, but in July 2012 the service was discontinued. Since the college culture was such that the majority of students were still traditional residential undergraduates and the library budget had not increased, the research team decided to implement Zoho Chat, another free online chat resource, until a spike in online research chat statistics merited a paid chat service, such as Library H3LP. With the new veterans scholar online degree program that is slated to begin August 2012, the research team anticipates that there will be a greater numbers of students utilizing the online chat service.

Though LibGuides and an online chat service certainly allowed the library to connect with more students in a virtual context, the Ensor LRC still had to bridge certain technological gaps when working with students in a traditional face-to-face reference interview. Even though the Ensor LRC had many public computers available and students were carrying around laptops, students would often approach the research help desk with their smartphones and tablet devices. On these devices, students would bring up links to our online databases via an open web search or an article link sent to them through e-mail. Though research on a PC or laptop may seem preferable to a librarian over research on 5-inch (and smaller) smartphone screen, it didn't change the fact that students were researching using their smartphones and tablets. This evolution in technology and information-seeking behavior changed the way in which the research team needed to approach and work with students. Specifically, the research team set out to make sure that patrons were aware of the free mobile resources available as part of the library's subscription databases.

Even though many subscription library databases had free apps available on both Apple and Android platforms, our research team had not made a large-scale or campuswide effort to advertise these resources. Hence, we worked to inform patrons on how to access these free apps on their smartphones and tablets to improve the research experience via mobile-friendly interfaces provided by our database vendors and apps that the Ensor LRC has developed. Of the subscription library databases, our research team identified and promoted the following apps and mobile interfaces for patrons to use on their smartphones and tablets: American Chemical Society database (app, free for download), BioOne database (mobile interface), Ebsco databases (app, free for download), Lexis-Nexis (beta mobile interface), LRC (app created by the Ensor LRC, free for download), and the LRC Catalog (app created by the Ensor LRC, free for download).

MOBILE PROJECTS AT THE ENSOR LRC

In addition to marketing existing resources, the Ensor LRC has undertaken various projects to make the library mobile friendly. Those projects include the purchase of iPods, the proposal and purchase of e-readers, and the development of additional mobile resources, two of which have already been mentioned (the LRC and LRC Catalog apps). The question is, as the title of this chapter implies, can you really make the library mobile on a shoestring budget? That is an important question considering the state of our economy, but a more important one may be, can you afford not to make it mobile? Statistics show that the use of mobile devices is growing, and libraries must try to meet the demands of their patrons if we as libraries and librarians are going to compete and survive in the world of giants such as Amazon.com.

Costs can be quite low if you have someone on staff with the skills to initiate and implement mobile projects and resources.

The first purchase was for two 8-GB Apple iPods to use with Shelflister, a Voyager inventory program. Library student assistants use these to do shelf-reading assignments, and it makes the task much more enjoyable for them and improves accuracy. The cost was $378 for both. (Shelflister is discussed in more detail later in this chapter.)

Additionally, most database providers include mobile options with their product without additional cost. EBSCO, for example, once charged an annual fee of $250 for a product called Adobe Content Server, which was needed to make the EBSCO e-books downloadable. Now, EBSCO waives that charge and provides that product for free.

As emerging technologies change the ways that people read and research and as we move from an environment of primarily print resources to electric ones, the Ensor LRC wanted to explore ways to make more patrons aware of available e-resources and offer something new and fun to them in the way of e-readers. We had all experienced the somewhat inconvenient and unpleasant reading experience that comes with reading e-books on the computer screen. So, we were excited about the notion of supplying our faculty, staff, and students with e-readers that could be checked out to improve their e-book reading experience and make the e-books portable. Purchasing e-readers for circulation would make the more than 140,000 e-books in our current collection much more accessible and create a better reading experience. It would also allow us to put multiple copies of both scholarly and popular works into circulation at a cheaper cost. Our proposal was to purchase six Kindles and six Nooks along with cases for each. We planned to transition a small portion of our current funding for print books to e-books. This was the first time this had been allowed. Due to accounting rules put in place years before people even thought of e-books, we were restricted to purchasing only physical print books with our book budget. The account number for which we pay for books actually means something in accounting terms. It means that purchases made from that account are fixed assets. As e-books are becoming more familiar, we are now able to justify their purchase as fixed assets as long as we purchase titles with perpetual access. This is still a new concept, and we are not the only ones having a hard time adjusting in the academic world. There are still unresolved issues of platform fees combined with perpetual title purchases, sales tax issues, workflow issues, and issues with digital rights management. Despite these issues, we moved forward with our proposal, and it was approved.

On July 25, 2012, the Ensor LRC began circulating six Kindles and six Nooks. This was an investment of $2,712, including one Kindle, five Kindle Keyboard 3G models, six Nook Simple Touch models, 12 protective cases, and

$1,000 in e-book content. The Amazon Kindle and the Barnes & Noble Nook are the most popular e-readers on the market, and the Ensor LRC already had accounts with both companies, so that made those models attractive.

You may be asking yourself, why the one Kindle and five Kindle Keyboard models? We first wanted to purchase the cheapest Kindle model, and so we ordered just the one to try it out—and luckily so! We quickly realized that it wouldn't work well with our campus wi-fi. Registering the Kindle device is practically impossible without a wi-fi connection, as is the downloading of content. A couple staff members of the Ensor LRC had the Kindle Keyboard model and liked them, so we decided that would be the model we would go with for the remaining purchases. The rationale for purchasing six of each is that e-book content can be shared among six devices per account; so, in essence, you are getting six copies of a book for the price of one. These e-readers can also access thousands of free titles because most of classic litera-ture published prior to 1923 is in the public domain. This makes the devices a very cost-effective way to keep classic works in our library's holdings.

The Ensor LRC has mostly purchased content for the Kindles, since the idea was to check those out with preloaded popular fiction titles. Titles were chosen on the basis of the best-selling titles according to the 2011 *Library and Book Trade Almanac* along with patron suggestions. Our plan is to add free titles as patrons request them. We do not want patrons downloading titles to the Kindles themselves. Very few titles were purchased for the Nooks, but suggestions will be taken for future title purchases. The plan for the Nooks was different from the Kindles. The idea for the Nooks was to use them for downloading EBSCO and eBrary e-books. We hope that this will serve as a means to market our e-books and encourage patrons to use our e-books, since they will be easier to read on the Nook.

Back to our discussion on budget, we used our supply budget to purchase the devices and cases and our book budget for the content. Tables 4.4 and 4.5 list some pros and cons for the Kindle and the Nook.

You can keep costs at a minimum by using the mobile options provided by database providers and your integrated library system. You can seek out products such as the Adobe Content Server from EBSCO to make your e-book collection downloadable. You can redistribute funds typically reserved for print materials and reallocate them toward e-books, e-readers, and other electronic resources. Additionally, if you have the knowledge within your staff, you can develop additional mobile resources, which is exactly what the Ensor LRC was able to do.

The first mobile resource that was developed at Ensor LRC was a mobile version of our Voyager catalog.[2] However, rather than trying to develop something from scratch, we relied on examples and coding from other li-

Table 4.4.   The Kindle: Pros and Cons

Pros	*Cons*
• Ease of operation. • Easy to purchase. • With wi-fi or 3G connection—easy to download Kindle books. • Lots of free Kindle books. • Can download content to six devices.	• Amazon's 1-Click payment method can be a challenge for libraries—can only purchase one title at a time. • Must deregister Kindles before circulating them so that patrons can't purchase content; other options include taking out credit card on the Amazon account and/or possibly using parental controls to block purchase of content. • Tax—must have tax-exempt certificate on file with Amazon for Amazon Digital Services. • E-books sold by sellers other than Amazon will accrue tax, and each seller has to be contacted individually for a tax refund.

Table 4.5.   The Nook: Pros and Cons

Pros	*Cons*
• Ease of operation. • Easy to purchase. • With wi-fi connection, purchased content automatically shows up on device upon registering it.[a] • Touch screen is nice. • Can download content to six devices. • Password protection for purchasing content with device, so you don't have to deregister the device to circulate. • Supports download of EBSCO e-book and eBrary e-book content. • Supports ePUB and PDF, so you can save EBSCO journal articles or other PDFs to the device. • Can transfer books on Google Play to Nook.	• Can only purchase one title at a time. • If you deregister device, the purchased content will be erased. • Tax charged.

a. Supports WPA and 802.11 b, g, and n.

brarians within the Voyager community. After looking at various examples and experimenting with some of the coding available, we released a mobile version of the library catalog on November 14, 2011. The first release of the mobile catalog resembled the interface of the full catalog, just scaled down to fit on the screen of a mobile device. Through subsequent updates, the mobile catalog has been enhanced using the jQuery Mobile framework, which is a touch-optimized framework that has been developed to be compatible with a wide variety of smartphone and tablet platforms.

Also in connection with our Voyager system was the implementation of the Shelflister program. Shelflister is an inventory program that was written by Michael Doran, systems librarian at the University of Texas at Arlington. The interface is designed for a mobile device, and as a result, the Ensor LRC purchased two iPods to use with this program. Prior to implementation, student workers would do inventory by shelf reading with a clipboard and a pen. This seemed to be good alternative and one that students would enjoy more.

To begin, student workers enter a starting barcode and an ending barcode (Library of Congress call numbers can also be entered). Once the starting and ending barcodes are entered, a list of items, by either call number or title, is generated on the basis of what items should appear within the range of the barcodes entered. If an item is missing, damaged, or misplaced or falls under any other category that is listed, student workers can mark it under the relevant category. Once an item is marked within a certain category, a text file is created on our Voyager server. Unfortunately, the program does not automatically update Voyager if an item is marked as missing, damaged, or under any other category. Once the text file is created, the information can then be put it into a user-friendly format that can be used for the circulation staff to go in and manually update the records. The code for the program is available on Michael Doran's site,[3] which has a demo of the program as well. Shelflister was fairly easy to install, and there were just a few configurations that were required to get it up and running.

In addition to developing a mobile catalog and implementing Shelflister, the Ensor LRC revamped the interface of the online catalog and, in the process, wanted to add some mobile-friendly features. To accomplish that goal, two QR code features were added to the catalog. One feature added was "Get Info in a QR Code." Once users click that link, a window pops up with a QR code that, once scanned, displays the title, call number, and location of the item. The second feature added was the display of a QR code. Once users scan this QR code on their mobile device, it takes them to the webpage of the record. An additional webpage was created to explain what QR codes are, what is needed to read them, and how we are using them in our catalog. Patrons can access that page by clicking on the "What's this?" link right below the QR code. The coding and instructions on how to add QR codes to a Voyager catalog are available on the EL Commons Codeshare website.[4]

The Ensor LRC also decided to develop an iPhone application. That decision was based on information obtained from the information technology services department showing that more than 800 of the registered mobile devices on our campus wi-fi network were iOS devices. The LRC app[5] was developed with jQuery Mobile and PhoneGap. PhoneGap is an HTML5 platform that allows you to use existing web technologies, such as HTML, CSS, and JavaScript, to build native-like applications that can be submitted to the App Store. PhoneGap can be used to develop for any of the other major mobile platforms, such as Android and Blackberry.

The features included in the app are library hours, a catalog search, and a research section that includes links to all the accessible mobile databases. Additionally, the research section includes available apps from library vendors, such as the American Chemical Society, EBSCOhost, and NAXOS Music Library. There are YouTube links to tutorials from Credo Reference, EBSCOhost, and JSTOR. The Ask-A-Librarian section has features that enable users to e-mail a librarian, call a librarian, and proceed to the directory. A campus map and library maps are included in the app as well.

In conjunction with the LRC app, a second mobile catalog was developed, which is just a different skin file on the Voyager server. The appearance of the second mobile catalog is identical to the other mobile catalog. The reason that an additional mobile catalog was created was to track how many users are accessing it from the app. The statistics have shown that most of our users are accessing the mobile catalog from the app and not through a mobile browser. As a result, the library released a mobile catalog iPhone application: LRC Catalog.[6] The LRC Catalog app has a built-in barcode scanner that interprets ISBN barcodes and QR codes. This app was created with the coding made available by the Ryerson University Library.[7]

The latest mobile project that the Ensor LRC completed was a mobile website[8] in conjunction with the redesign of the library's full website. Like the mobile catalog and the LRC app, the jQuery Mobile framework was used in the development of the mobile website, which includes many of the features that are available in the LRC app, although there are some subtle differences. There are some design differences, and the layout of the information varies between the app and the mobile website for users to distinguish between the two. One distinct difference is in the research page of the mobile website. In that page, there is a link for apps, as there is in the LRC app, but when patrons click that link, a pop-up window appears prompting them to choose whether they want to be directed to a page for available Android or iOS apps. Despite the differences between the app and the mobile website, patrons will have an equal yet unique experience with each.

CONCLUSION

As more and more of our users become equipped with various mobile devices, it is imperative that we begin looking at ways to provide mobile-friendly access to the library's resources. It has become all the more important now since smartphones and other mobile devices are outpacing the sale of personal computers (Graziano, 2012). It is true that shipments do not equal sales, but it is a telling sign of a trend that retailers are expecting to see—that is, more consumers purchasing smartphones or other mobile devices over traditional computers. What this means for libraries is that no longer are patrons simply going to want online access; they will want and expect mobile access. As a result of the expansion of mobile technology, the Ensor LRC has to create a mobile presence for the library. That has been done through the purchase of iPods, e-readers, and the development of different mobile resources. The result of these initiatives has been the positive feedback that we have received from our campus community and the usage that we have seen with the mobile resources, particularly the library apps. However, our work is far from over. As mobile technology continues to grow, the Ensor LRC plans to consistently evaluate its mobile presence to ensure that the library is where it needs to be. The idea of creating a mobile presence may seem a little daunting at first. The best advice would be to start with what you have available to you already through subscription databases. Many vendors are developing either mobile-optimized interfaces or apps (some are doing both). Although it may seem daunting, remember that our patrons are going to expect mobile access. Library vendors are realizing that, and we as libraries and librarians need to as well.

NOTES

1. Georgetown College is a private residential liberal arts college in Georgetown, Kentucky. The curriculum includes traditional face-to-face undergraduate classes and online programs in the graduate education and veteran scholars programs. The Anna Ashcraft Ensor Learning Resource Center serves a full-time equivalent of 1,661 students and 198 full-time staff, 53 part-time staff, 116 full-time faculty members, and 45 adjunct faculty members.

2. The link to the Ensor Learning Resource Center's mobile catalog is https://voyager.georgetowncollege.edu/vwebv/searchBasic?sk=mobile.

3. http://rocky.uta.edu/doran/shelflister/.

4. The link to the instructions and code is http://www.exlibrisgroup.org/display/VoyagerCC/Adding+QR+Codes.

5. http://itunes.apple.com/us/app/lrc/id497416254?mt=8.

6. http://itunes.apple.com/us/app/lrc-catalog/id541057939?mt=8.

7. The link to the code from Ryerson University Library is available in the Code4Lib article "ISBN and QR Barcode Scanning Mobile App for Libraries," http://journal.code4lib.org/articles/5014.

8. http://www.georgetowncollege.edu/library/mobile/.

REFERENCES

Georgetown College. (2011). *President's report to the Board of Trustees*. George-town, KY: Author.

Graziano, D. (2012). *Canalys: Smartphone shipments surpassed PC shipments in 2011*. Retrieved from http://www.bgr.com/2012/02/03/canalys-smartphone-shipments-surpassed-pc-shipments-in-2011/

Raine, L. (2012). *Tablet and e-book reader ownership nearly double over the holi-day gift-giving period*. Retrieved from http://libraries.pewinternet.org/2012/01/23/tablet-and-e-book-reader-ownership-nearly-double-over-the-holiday-gift-giving-period/

Smith, A. (2012). *Cell Internet use 2012*. Retrieved from http://pewinternet.org/Reports/2012/Cell-Internet-Use-2012.aspx

5

The Orange County Library System Shake It! App

Cassandra Shivers
Orange County Library System

ENGAGING OUR COMMUNITY THROUGH AN APP

The Orange County Library System (OCLS; Florida) Shake It! app for iOS (Apple iPhone, iPod touch, iPad) was made available to the public through the Apple App Store in July 2010, a year after development on the app began. The idea of creating an iOS application began percolating in the information systems department in early 2009 after staff at OCLS began using Apple iPod touches for the first time, along with a variety of applications available through the App Store. The idea lingered while we assessed what would be required to create the app and if it would be possible to do in-house. Development began in earnest June 2009 after it was determined what the app would do and that our web design specialist could learn the skills needed to create the app interface and functionality.

The OCLS continually seeks new and innovative ways to engage with the community. With the rise in popularity of smartphones and mobile applications, it made sense for the organization to step forward with application development and to reach out to its patrons in an unexpected manner through a mobile platform. If we're able to connect with our community members in surprising ways, it's an opportunity to change their perception of libraries as book warehouses or information repositories. If we can add value to their daily lives and meet them on the go, beyond the walls of the library, we increase the OCLS's relevancy in the everyday lives of our patrons. Creating a mobile application that encourages library material discovery and use of the catalog is a venture in that direction.

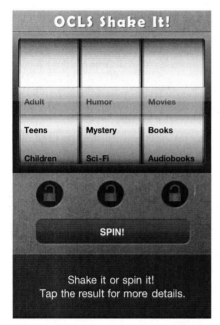

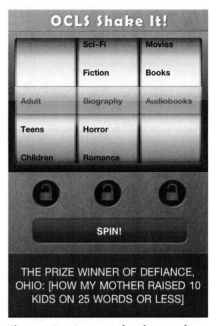

Figure 5.1. The home screen of the Shake It! app.

Figure 5.2. An example of a match result for an adult–biography–audiobooks combination.

THE OCLS SHAKE IT! APP

Patrons shake their iPod touch, iPhone, or iPad, and the OCLS Shake It! app returns a title from our catalog (Figure 5.1). The app utilizes a "slot machine" interface, and shaking the device or tapping a "spin" button causes the slot machine spinners to rotate before settling on a combination, which is then displayed, and a title from the library collection that fits that combination match is presented (Figure 5.2). The resulting title can then be tapped to launch the mobile catalog, from which details about the title can be viewed, including location, call number, and status. If the item is available for home delivery, it can be also be requested at that time.

HOW IT WORKS

There are three selectors, or "spinners," in the slot machine, one each for audience, genre, and format. Each spinner can be locked down to return a specific type of random title or left unlocked for a truly random selection from the library catalog. The audience spinner consists of adult, teens, and children. Genre is broken down into humor, mystery, sci-fi, fiction, biography, horror, and romance. Movies, books, and audiobooks make up the options in the format spinner. When deciding on the genres, we wanted options that worked across format; for instance, the humor genre works just as well for humorous books and audiobooks as it does for motion picture and television comedies. We briefly considered including music CDs as a format, but that material type didn't lend itself to the genre selector. Originally, we had a few more nonfiction genres but decided to focus the app on fiction and biographies, as the app was intended for superficial searching of the catalog and not in-depth research or advanced searching.

Results in the app based on audience are enforced mostly through call number prefix, with children's material starting with "Children" or "Family," teen material beginning with "YA," and all other call number prefixes qualifying as adult material. It wasn't possible to use this logic when determining movie results for the teen audience, so we incorporated the MARC 521 field. In essence, if the MPAA rating was present and listed with the bibliographic record, it could be used to make exclusion statements. Specifically, movie titles with ratings higher than PG-13 wouldn't come up as a result for the teen audience; that audience would be served movies with ratings of G, PG, or PG-13. This seemed like a practical solution, as anyone interested in accessing materials with these ratings would have an avenue to do so; anyone seeking materials with higher ratings could also be served by locking down the adult audience, whether they were adults or teens in actuality.

THE APPLICATION DEVELOPMENT PROCESS

Development of the OCLS Shake It! app began with the OCLS web design specialist learning to use the software for writing the necessary code for the application. This free software included Xcode and the iOS SDK (development kit). To create the app, he needed to know basic coding principles, such as functions, variables, and loops. He also learned aspects of Objective-C. During this time, we also purchased a Mac-Mini, Apple Developer License, and an iPod touch, all necessary tools in iOS application development.

Once the app interface and functions were created, our web developer (who is fluent in seven programming languages) began writing the code that would connect the app to a database of catalog records. The Shake It! app uses JavaScript on ASP, mySQL, PHP, HTML, and XML. He subsequently wrote the script that queries the database according to the input received from the app. To make the connection between the app selectors (audience–genre–format) and the random resulting title from the catalog, the database queries had to be established upon predetermined catalog search results. One of the OCLS digital access architects worked on providing the web developer with this criterion.

There are 63 possible shake combinations in the app. For every possible combination, a strategy had to be determined to pull a title from our catalog matching the given slot machine combination. Fortunately, most of the matches could be determined by call number, since that often includes audience, genre, and format. For example, a title starting with the call number "J CDB FIC" will match the spinner combination of children–fiction–audiobook. To return more specific results that match a subgenre of fiction, other fields of the bibliographic record are considered, including the 650 subject field and the 655 genre field. To have the app return a title for children–mystery–audiobook, the mySQL query looks for items with a call number that starts with "J CDB FIC" and has the word "mystery" or "detective" in the 650 or 655 field of the MARC record in the associated bib record of a title; then, from the list of matches, it returns a random result to display in the app interface.

We built the app in such a way that all possible combinations always return a result (unless there's a network connectivity issue between the app and database). This required creativity in the match logic when there are random title matches for combinations such as children–romance–book and children–horror–movie. In addition to ensuring a result for every shake, we've grossly eliminated some catalog records from our database of records that the app pulls from. We consistently had an issue where the movie *A Clockwork Orange* was returning for the teens–sci-fi–movie combination. This was a result of the word "teenager" in the 655 field of the bibliographic record for *A Clockwork Orange*. Our intent was for use of the word "teenager" in the 650 or 655 field to help designate a title as fitting for the teen audience. Unlike children's movies, we can't simply rely on a call number prefix to identify audience-appropriate movie selections for teens, because we don't have a YA DVD–designated collection. The simplest solution in this case was to remove records for *A Clockwork Orange* from our database of records that the app pulls from.

In addition to the web design specialist, web developer, and digital access architect who worked on the creation of the OCLS Shake It! app, the digital content specialist, also in the information systems department of the OCLS, contributed the sound effects and music heard in the app as the slot machine spins, the locks are tapped, and the random title is returned. Roughly 130 staff hours were spent in the development of the iOS app. In the spring of 2011, the Shake It! app was successfully ported to the Android OS and released to the public in May 2011. The Android Shake It! app interface and functionality were developed using the Eclipse IDE by a University of Central Florida software engineering student who interned in the information systems department for a semester.

KEEPING THE SHAKE IT! RESULTS FRESH AND ACCURATE

New material is added to the library catalog every week. Many items don't circulate until their Tuesday release date, so the catalog export is run every Wednesday to ensure that new material is available to be returned as a random result of the app. Currently, the export process is manual and requires OCLS staff attention. Every week, over 200,000 catalog records are exported from the OCLS integrated library system that meet the following conditions: it was published after the year 2000 and currently has a copy available for checkout. Information for each bibliographic record that's exported includes bibliographic record number, material type, created date, call number, publication information, series, MARC 245|a, author, edition, MARC 650, MARC 655, MARC 521, and MARC 245|b. The export is saved as a TXT file, which is then imported into the mySQL database. Most of the fields in the database are used to determine which title to display after a shake. The contents of the bib record number field are used to form the direct link from the title in the app to the display of the title in the mobile catalog, which was created by a third party and provided through our integrated library system vendor, Innovative Interfaces. Parts of the MARC 245 field are used to form the text that displays the title of the match in the app (Figure 5.3).

Each time the export is run and imported into the database that the app runs against, everything is tested to ensure that the app is working properly and titles are displayed as expected for the various slot machine combinations. To aid staff in this weekly testing, the web developer created an internal web application that also runs off the database of bibliographic records and can be used on a PC. Using this internal application on a computer, staff have the ability to take the "randomness" out of the results and view a complete list of possible results from the database that match a given combination. The internal viewer application allows staff to select an option from each app spinner and submit the request to the database. What results is a list of hundreds of titles that match that criteria, any one of which could be one of the randomly returned titles in the Shake It! app. The titles that are listed in the internal viewer also link to the mobile catalog for easy checking of the title links. The internal viewer application is an essential tool in continued development of the Shake It! app (Figure 5.4).

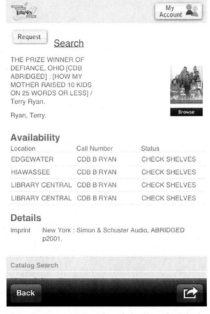

Figure 5.3.   Example of the result title in the mobile catalog.

SUCCESSFUL APP LAUNCH

Once the OCLS Shake It! app was approved in the Apple App Store and available to the public for download, the Library began promoting it in a variety of ways to the community. It was advertised on the web through the home page of the library website, the OCLS General News RSS feed, and two library e-newsletters (general and technology). It was promoted in print through the library's monthly newsletter, *Books & Beyond*, and posters throughout its 15 locations. An entertaining stop-motion video featuring "Doug & Sam" (G.I. Joe action figures routinely used in library video promotion) was also created featuring the OCLS Shake It! app (Figure 5.5).

In the first 6 weeks of its release, the Shake It! app was downloaded more than 2,000 times and shaken over 8,000 times. Early reviews in the App Store included such comments as "This is super convenient. Love it" and "Neat idea, glad to see OCLS w an app." In the 2 years since its launch, the OCLS Shake It! app has been downloaded more than 7,000 times and has been shaken over 130,000 times, averaging about 300 downloads a month and over 5,000 shakes a month. Through July 2012, the Android version of the app has been installed almost 1,700 times. On average, the Android OS version has

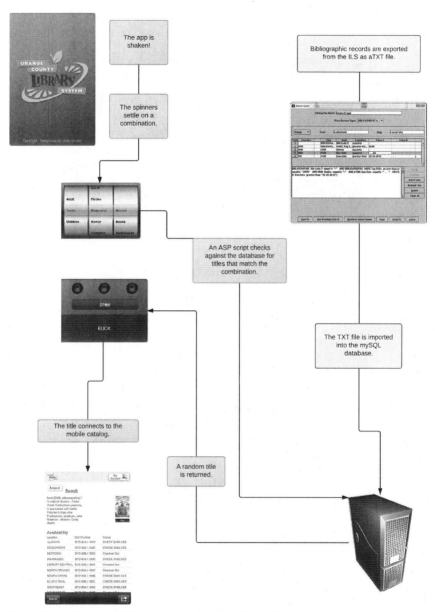

Figure 5.4. Diagram of the components of the Shake It! app.

Figure 5.5. Screenshot of Doug with the Shake It! app, from the "Doug & Sam" video.

been downloaded about 100 times a month. These statistics are an indication that the OCLS's goal to connect with the community and provide a fun and engaging way to discover the collection has been successful.

THE FUTURE OF SHAKE IT!

OCLS has many exciting plans for future versions of the app. The team has been inspired by an array of apps in the marketplace, as well as by feedback received from patrons. While the library's intention was to create a discovery tool and not a resource for studious research, it's clear that an avenue directly into the catalog, via the app, for a traditional search of the collection would be welcomed by the community. OCLS has also received multiple requests to have the slot machine match return more than one title at a time. Social media has become an even more prevalent means for communication in the time since the app was first launched, and the library would certainly like to integrate a method for users of OCLS Shake It! to be able to push their "shake" activity out to their preferred social network, including Facebook and Twitter. People using the app could share the results of their shake or the criteria they've locked down for their shake.

The use of virtual badges is a rising phenomenon that would truly lend itself to an application like OCLS Shake It! The possibilities are endless for how a badge system could be implemented with the app that in turn would provide further incentive and result in prolonged use of the app. Users could earn a badge when they download the app for the first time, spin the slot machine for the first time, or share their app activity through social media for the first time. Badges for each genre could be earned as various random combinations are returned or if a specific genre has been returned a given number of times.

The OCLS would also like to utilize a prompt to encourage review and rating of the app by its users. Such a prompt would be subtle and appear only after a designated amount of use of the app has occurred, displaying a limited number of times during the lifetime install of the app.

A more challenging revision to the app that the library would certainly like to accomplish is the inclusion of downloadable media into the app. In addition to the existing material types displayed in the format spinner, we'd like to list downloadable audiobooks and e-books as options. The OCLS has a large collection of downloadable media, and OCLS Shake It! would be a marvelous fit for helping patrons discover this wonderful collection. Our plan is to only match titles with file formats of audiobooks and e-books that are compatible for direct download through iOS and Android OS devices, such as MP3 and EPUB. Our goal is for the experience to be seamless so that if patrons are shown a downloadable audiobook title through the Shake It! app that they'd like to listen to, they can complete the checkout and download process entirely through their smartphone or tablet. This is the first part of downloadable media integration, which is challenging. The majority of the thousands of bibliographic records we have for downloadable media in our integrated library system don't contain specific file format information—that is, whether the title is available in WMA, MP3, EPUB, or PDF. At this time, the expense to update these records is prohibitive. The other major challenge for including downloadable media in the app concerns the strategy for determining which titles result with the slot machine combinations. Because downloadable media don't have shelf locations, they don't have call numbers. Call numbers are the foundation through which the majority of the audience and genre matches are determined in the app. While integrating downloadable media into the OCLS Shake It! app appears daunting, due to the perfect fit of the delivery system (the mobile app) and the media (downloadable), the effort finding a solution to the challenge is well worth it.

KEYS TO BUILDING AN APPLICATION FOR YOUR LIBRARY

The OCLS Shake It! app is a terrific example of what can be created in-house by a library. It all begins with an idea that addresses the question, what service, product, or collection does your library have for which you want to engage your community through a mobile delivery system? The app you create can be a simple web app, essentially a webpage or website that is formatted for use on mobile devices and viewed through the device browser; or, it can be a native app that makes use of the special functions of mobile devices, such as the accelerometer, compass, and camera. A native app is downloaded from an app store and installed on the device to be used. Web apps can be updated on the fly, with updates received by the users the very next time they access the page on their device. Native apps have to be submitted to the app store and downloaded from there before any updates are received. There are many factors to consider when determining which type of app you want to create for your library.

There may be a few obstacles to overcome in developing an app. One of the biggest challenges may be in finding staff with the skill set to create and develop the app. If there isn't someone on staff who knows how to code applications or who can be given the time to learn, consider bringing in an intern from a nearby college or university who is studying computer programming or mobile development. It's a great opportunity for a student to get practical experience in a real-world environment. Often libraries don't have direct access to their data; for instance, if an organization wanted to have an app that worked with its catalog or event calendar, it may be tricky obtaining the content. It won't hurt to ask for access to your data, and you may find that your vendor provides application programming interfaces that can be used to deliver your content through a mobile platform.

It can be fairly inexpensive to build an app, cheap even, if you create a web app on existing server space. If you rely on another organization for your website, try asking for some space on the server for mobile development. Or consider using an external web hosting site. If you are developing for iOS, you'll need a Mac computer for development, and at the point you're ready to test your app on a real device (and not the simulator provided through Xcode), you'll need an Apple Developer license, which is $99 a year. You can purchase a used or refurbished iOS device for testing. Adobe DreamWeaver and Photoshop are recommended tools for coding and graphics creation. They're the industry standards, and as such, you'll easily be able to find help through forums and tutorials on how to use them. There are less expensive and free alternatives, such as Komposer for coding and Gimp for image editing.

There are many great resources about application development, not just in print but on the web. Anyone shouldering the responsibility of native app or web app development for their library can find valuable information simply by searching Google for the key terms of the particular hurdle they are trying to overcome in the development process. Chances are, someone else has faced the same obstacle, and through the process of shared troubleshooting, the development community has generated a solid knowledge base on the Internet supporting app development.

6

The North Carolina State University Libraries' Mobile Scavenger Hunt: A Case Study

ANNE BURKE, ADRIENNE LAI, AND ADAM ROGERS
North Carolina State University Libraries

With over 34,000 students, North Carolina State University is the largest 4-year institution in North Carolina. Over 4,500 first-year students arrive on campus each year, and most of them have had neither the cause nor the opportunity to use a large university library. It is therefore not surprising that many students are overwhelmed by the size of D. H. Hill, the main library at North Carolina State University.

The North Carolina State University Libraries' best opportunity to reach students early in their college experience and teach them how to use the library to their advantage is through ENG 101: Academic Writing and Research. Teaching students how to locate, analyze, synthesize, and use information is a key component of ENG 101, and the libraries have traditionally supported this objective through face-to-face instruction and online tutorials.

Prior to the launch of this scavenger hunt, library orientation typically took place in a computer lab, with a librarian telling students about the libraries and showing them around the website. Finding little immediate application for this information, students often tuned these sessions out. Furthermore, classroom-based sessions did not address the one thing that the students whom we surveyed reported being most confused about: how to navigate our often-confusing 220,000-square-foot building.

In addition to surveying students about their library instruction experience, we interviewed First Year Writing Program (ENG 101) faculty to find out how the library could better support their teaching. The ENG 101 instructors whom we spoke to focused on the library anxiety factor. "Students are very, very hesitant to ask [librarians] things," one faculty member noted. Another remarked that most of her students admit to never having been to any brick-and-mortar library, much less one of our size. The faculty members with whom we spoke seemed to agree that student library anxiety would be largely ameliorated by more face time with a librarian but that the library sessions should be fun, active, and engaging.

PROJECT ORIGINS

In 2008, two North Carolina State University librarians had experimented with a library scavenger hunt as an alternative to the classroom orientation model. This paper-based scavenger hunt was successful, but the task of hiding paper clues throughout the library was too labor intensive to scale the activity up beyond a handful of offerings a year; it was definitely not sustainable enough to meet the needs of over 100 sections of ENG 101 each semester.

By 2011, the growth of mobile devices, cloud computing, and online collaboration tools seemed to offer a solution to the scalability issues of a scavenger hunt–type activity. These tools would allow the exploration of physical and virtual library spaces through a combination of game dynamics, mobile technologies, and experiential learning. For example, as part of its 100th-anniversary celebrations, the New York Public Library (2012) developed "Find the Future," a game that led participants on an exploration of the library's collections and spaces via a custom-built smartphone app. This highly publicized game was an inspirational model for our project.

In March 2011, we started exploring how we could create a similar mobile device–enabled game to introduce students to our library spaces, collections, and services. Our primary learning objectives for this activity were typical library orientation goals: to introduce students to library resources and services, to have students use the library website to find books and journals, and to orient students to the library's physical spaces. This was also an opportunity to introduce students to mobile, cloud-based technologies in an academic setting. We also envisioned a number of affective goals for this activity. We wanted to foster student confidence in using library resources and spaces; promote some of the cool, unusual, or most-used resources in the library; and convey the message that the library is a place for exploration and fun. After a couple of months of designing and testing, we ran a pilot of the activity in June 2011, offered it to ENG 101 faculty in fall 2011, and, by the end of spring 2012, had facilitated 90 scavenger hunts for almost 2,000 students.

RESEARCHING TECHNOLOGIES

Our informal project team began meeting regularly in the spring of 2011. We considered different technical solutions, apps, and approaches to mobile scavenger hunts. Early on, we decided that we couldn't rely on students having their own smartphones and would need to use library devices for the activity. The iPod Touch was an obvious choice because we already had access to several through our technology lending program, and it offered impressive capabilities at a reasonable cost. The fourth-generation model had been recently released, with wireless networking, a web browser, access to thousands of apps, and (for the first time) a camera.

At the time, Foursquare and SCVNGR were making news for their use of location- and game-based interactivity, so we investigated their apps. SCVNGR in particular seemed to fit our needs: it was made for scavenger hunts, ran on iOS, and was free for users. We set up SCVNGR accounts, installed the app on some iPods, and started making test "challenges." However, we very quickly ran into two obstacles: First, challenges were limited with a free account, so we would have to pay for an account to build a full prototype of our scavenger hunt. Second, the platform was focused on GPS-based check-ins, and our scavenger hunt would take place indoors, in a large building—effectively a single location. We tried workarounds for these issues, but it became simpler to move on to other options.

FOCUSING ON FUNCTIONAL REQUIREMENTS

We had started our research hoping to find a single app to meet our needs, but we gradually realized that this was an unlikely possibility. So we decided to identify the functional components that we needed to run the activity:

- A list of clues/questions
- A way to submit answers (as text, photo, possibly audio/video)
- A way to keep score
- Access to a map of the library
- Some way to help students through technical issues

And, of course, we wanted all of this for free: we were unable to spend money on apps or on a service with a monthly fee. Looking at the list, we realized that we were asking quite a lot of an app. So we prioritized: what were the most important features to have on the mobile device?

We wanted the students to engage with different kinds of media (websites, photos, audio/video) and to submit answers as text or images, so we focused on using the devices for submitting answers. We set out to find a mobile app for sharing text documents and photographs and found that the multimedia note-taking app Evernote suited our purposes.

For the activity as a whole, we ended up with a hybrid approach, combining the iPods and free iOS apps with web apps, paper documents, and human effort. In our design, only some of the scavenger hunt is done on the iPods: students use Evernote to submit answers and other apps to find answers. The list of clues, library map, and hints are all given to students as separate paper documents, and scorekeeping is done by library staff using the web (or iPad) versions of Evernote and Google Docs.

HOW IT WORKS

We typically run a scavenger hunt for classes of 20 students, which we divide into five teams of four. The classes meet in our instruction lab in the library, and the activity proceeds as follows.

Packet: Each team is given a packet with a clue sheet (some clues are the same; others are different for each team), a map of the library, a hint sheet, and an iPod Touch with certain apps installed and as logged into a specific Evernote for the team. We briefly teach the students how to use Evernote, set their iPod timers for 25 minutes, and send them out into the library.

Clue: Teams strategize which clues to answer in what order, following each clue's instructions to find a particular space or resource in the library or the answer to a question.

Answer: For each clue, the team submits an answer on the iPod by making a new note in Evernote, titling it with the clue number, and either typing in text or attaching a photo to the note. When the note is saved, it is uploaded to the Evernote servers.

Sync: Each team's Evernote account has been set up to share its notebook (the team's answers) with a master Evernote account, which library staff use to access the team's answers.

Score: One library staff member monitors the master Evernote account, looking for newly posted answers. When an answer comes in, she or he makes a note in a Google Doc spreadsheet, counting that answer as right or wrong for that team. The spreadsheet has been previously configured to automatically tabulate points for answers marked correct.

Show: Meanwhile, another library staff member separately copies students' photos from the master Evernote account into a PowerPoint template, which is shown to students (on a loop) when they return to the instruction lab. We ask students to complete a short survey for assessment, and we address their questions. When the scorekeeper has finished, we announce the winning team and give out prizes.

THE PITCH

When we had a viable prototype of the mobile scavenger hunt, we tested it out with our student workers, made some refinements, and then developed a "sales pitch" for the project. We strategized about partnerships for the short and long term and decided to pitch the idea to North Carolina State University's First Year College (a living and learning community for undeclared first-year students). It had worked with the library on other projects, was committed to its students learning library skills, and was open to new ideas. In our pitch to First Year College instructors, we showed photos of our student workers doing the scavenger hunt, talked about our goals and learning objectives, and asked for feedback. The instructors were enthusiastic, and one promptly asked us to run a scavenger hunt for a summer class just a few weeks later.

THE PILOT

We approached this first class as a pilot, explaining to the students that we were trying something new and looking for feedback. The class was an upper-level communications class, so while they were not our target audience of first-year students, they made for a fine test audience. We mostly followed the format described earlier, with very positive results. In feedback forms, the students nearly all reported that they enjoyed the activity, found the technology simple to use, and learned something new about the library.

We encountered no major technical or logistical problems in our scavenger hunt pilot but found many things to improve. We noticed that it was very difficult for the scorekeepers (we used two) to match the pace of the teams' answers. We decided to lower the number of teams (from seven in the pilot to five in our current practice), which solved this and allowed us to use only one scorekeeper. The students enjoyed seeing their photos at the end (we had put them up in PowerPoint on a whim), so we decided to make that a standard part of the activity. We improved many other things (e.g., our explanation of Evernote), and we have continually refined the activity as issues or opportunities have arisen. To share the results of the pilot with our supervisors, we wrote up a report, including student feedback and photos.

RESOURCES

One of the major appeals of this activity is that it is fairly inexpensive. To run the pilot, we used iPod Touches from our existing technology lending collection and the free versions of all the apps. Once the pilot was a demonstrated success, we were able to purchase 10 iPods to dedicate to the activity ($199/each.) We seldom use more than 5 iPods in any one session, but 10 iPods are necessary when running several consecutive scavenger hunts. We use a laptop and an iPad for scoring, but this could all be accomplished on just one computer. We distribute prizes to the winning team and consolation prizes to everyone else. The cost of this was under $500 for 90 scavenger hunts with over 1,800 students.

The scavenger hunt is most costly in terms of staff resources; each hunt consumes about 120 minutes of staff time: 10 minutes of setup, 50 minutes of active class time (with two staff), and 10 minutes of cleanup. Two library staff members are needed for the scavenger hunt to run smoothly: an emcee and a scorekeeper. Initially, we had two professional librarians scheduled to every scavenger hunt. We have since, however, been scheduling one librarian (emcee) and an undergraduate or graduate student assistant (scorekeeper).

USER COMMUNITY REACTION

Since launching the project, we have had consistently positive feedback from our users. The First Year Writing Program faculty responded very enthusiastically from the start: for the first semester that we offered the scavenger hunt as an option for library orientation, instructors for 45 classes requested it (out of over 100 per semester). Some asked for a scavenger hunt in addition to a traditional library instruction session. The overall result was a dramatic increase in the number of sessions that the library facilitated for ENG 101 classes. In the four semesters prior to the launch of the scavenger hunt program, librarians worked with ENG 101 classes an average of 21 times per semester (see Table 6.1). In the two semesters since the scavenger hunt program launched, librarians have interacted with ENG 101 classes an average of 71 times per semester.

Perhaps more important, the students have reacted to the scavenger hunts with great enthusiasm. We often witness a dramatic transformation in students' body language from when they enter the session (thinking they are going to spend the next hour in a dim computer lab listening to a librarian) to when they return to the room after doing the scavenger hunt. Instead of having the usual glazed, half-awake expressions typical of undergraduates at computer terminals, students are animated, engaged, and occasionally breathless.

The way we have structured the scavenger hunt activity offers some more opportunities for librarians to interact with the students in the informal, non-intimidating ways that the ENG 101 faculty recommended for combating students' library anxiety. We encourage students to be creative with their

Table 6.1. Number of Library Instruction Sessions and Scavenger Hunts Offered in ENG 101 Classes over a Six-Semester Period

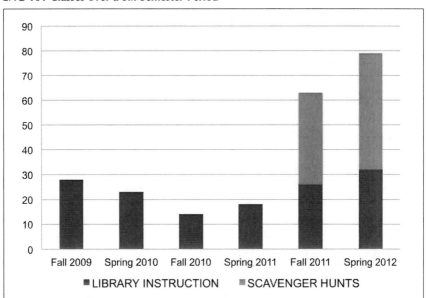

photograph answers (Figure 6.1), so they often have fun with these: rejoicing when they've found a book in the stacks, "planking" (Wikipedia, 2012) on some of the library's more unusual chairs, or "Tebowing" (Tebowing, 2012) with a reference librarian. Because we use cloud-based tools that allow us to monitor students' answers in real time, we can offer instant feedback. When they return at the end of the session, we can chat with them about the photographs they have taken or give an impromptu tutorial for questions that they answered incorrectly.

Figure 6.1. Photographs taken by students during the scavenger hunt.

ASSESSMENT AND FEEDBACK

Because this activity was new for us, we were very interested in learning what students and faculty thought. At the conclusion of each session, we asked students to complete a short online survey, and at the end of each semester, we sent another online survey to faculty. The surveys were created with Google Forms and Qualtrics. We were careful to keep both assessments brief to ensure maximum completion.

From students, we were keen to know how well the scavenger hunt met our affective goals. We wanted to know if they found the activity enjoyable, if they felt at ease interacting with library staff, and if they believed they learned something new. We also asked two questions regarding the scavenger hunt technology and logistics to make continuous improvements. In two semesters (fall 2011 and spring 2012), we surveyed 960 students, asking them to respond to the statements listed in Table 6.2.

We were also interested in learning how well faculty believed the scavenger hunt met its stated objectives, whether any other instructional objectives should have been addressed, and how satisfied they were with the communication and scheduling process. Between fall 2011 and spring 2012, 30 faculty members completed the survey (15 each semester; Tables 6.3 and 6.4). Because we changed the survey instrument and several of the questions between the semesters, each semester's data are outlined separately.

Anecdotally, we have noticed that the quality of the face-to-face interaction in scavenger hunt sessions seems to help break down the barriers between students and librarians. Ironically, although students actually spend less face-to-face time with the librarian during a scavenger hunt as compared to a traditional instruction session (students are sent out of the room for half the session in a scavenger hunt), the interaction tends to be more personal

Table 6.2.   Post–Scavenger Hunt Student Survey Data: Fall 2011 and Spring 2012 (in Percentages)

	Strongly Disagree	Disagree	Neutral	Agree	Strongly Agree
This was an enjoyable activity.	0.6	0.9	8.9	47.5	42.1
I learned something new about the library.	0.6	1.4	6.6	48.0	43.4
I feel comfortable asking a library staff member for help.	0.2	0.8	3.9	42.5	52.6
I found the technology simple to use.	0.3	1.3	8.1	51.0	38.3
I found the directions easy to follow.	0.6	0.9	8.9	51.5	38.1

Table 6.3. Faculty Scavenger Hunt Feedback: Fall 2011 (in Percentages)

	Very Poorly	Poorly	Neutral	Well	Very Well
How well do you believe the scavenger hunt met its objective to encourage problem based learning with team dynamics?	0.0	0.0	13.0	53.0	33.0
How well do you believe the scavenger hunt met its objective to decrease library related anxiety?	0.0	0.0	0.0	33.0	67.0
How well do you believe the scavenger hunt met its objective to increase awareness of library resources and services?	0.0	0.0	0.0	27.0	73.0
How well do you believe the scavenger hunt met its objective to aid students in navigating library spaces?	0.00	0.0	0.0	40.0	60.0
How well do you believe the scavenger hunt met its objective to introduce emerging mobile and cloud-based technologies?	0.0	0.0	13.0	47.0	40.0

and therefore more memorable for the students. It is not at all unusual for a student to see one of us in the library, smile, and say, "You did the scavenger hunt for my English class!"

The scavenger hunt's use of active and experiential learning not only makes the students' first library instruction session more personal and memorable but also seems to help students learn the basics of library resources,

Table 6.4. Faculty Scavenger Hunt Feedback: Spring 2012 (in Percentages)

	Yes	No
Do you believe the mobile scavenger hunt improved your students' awareness of library resources and services?	100	0
Do you believe the mobile scavenger hunt improved your students' confidence in using the library's collections and navigating its spaces?	100	0

Are there any other instructional objectives you would like to see addressed by the mobile scavenger hunt?

- "Perhaps an orientation to the main library webpage."
- "Perhaps more specific text-based items to locate."
- "Undergraduate-oriented resources/librarians. Undergrads often have trouble finding someone who can help them with basic research for core classes."
- "Scholarly v. popular sources? If that can be done. It might help reinforce that idea. Also, the idea that books can still be useful in research!"

services, and spaces. One ENG 101 instructor found that the scavenger hunt "led to increased understanding, deeper learning, and almost complete recall of important library functions" and that students reported that a subsequent assignment (to retrieve a book for an upcoming literature review) was less intimidating because of the mobile scavenger hunt. Student comments from the posthunt surveys seem to confirm this:

> "I think it's a little hard to navigate because of how big the library is, but it was a great way to start learning about the place!"
> "This was a good activity instead of just listening to someone tell us where everything is in the library."
> "I am surprised how many resources library has. And I will definitely ask . . . staff for help in the future."

By providing a low-stakes activity where students could—in the safety of a group—approach a librarian on the reference desk or venture into the stacks to find a particular book, the scavenger hunt allows students to be more confident when they need to undertake these same tasks in real-life situations.

Last but certainly not least, the scavenger hunt overwhelmingly fulfills our affective goals of improving student engagement with and sentiment about the library (Figure 6.2). Many of the student comments we receive in the posthunt surveys express a surprising amount of enthusiasm for what is, at essence, a library orientation:

> "I LOVED this activity! It made going [to] the library fun! I also got to meet new people in my class that I didn't already know. NC State's library is soooooo cool!"
> "It was AWESOME!!!!!!"
> "I wish we had more time to finish more of the questions."

Many libraries are seeking to convince students that the library is full of great learning spaces, valuable resources, cool technologies, and friendly, helpful staff. While it is one thing to tell the students how wonderful the library is, it is more effective to let them experience it firsthand.

LESSONS LEARNED

The use of free and consumer-level tools makes adopting the mobile scavenger hunt relatively simple, but following some best practices will ensure project success. As with any new project, start with goals and a plan for

Figure 6.2. Students answer a question during the mobile scavenger hunt.

measuring their accomplishment. As the project moves from an idea to a pilot and into a public launch, clearly communicate with project stakeholders to facilitate understanding about commitments and resource requirements.

Access to a handful of iPod Touches (or other mobile devices with similar capabilities, such as iPads or some Android tablets) is needed to embark on the mobile scavenger hunt project. Existing library or campus resources may provide these, such as a technology lending program or a campus educational technology grant. And although implementation of the mobile scavenger hunt doesn't require a programmer or technology wizard, someone in the project team will need to be tech savvy and comfortable configuring iPods, or some light information technology support will be needed.

It may go without saying that this activity is somewhat dependent on consistent wireless connectivity in the library where it takes place. Wireless doesn't need to be perfect everywhere—Evernote will cache notes locally and sync them when wireless is available—but it is important to make note of any known wireless infrastructural issues.

The mobile scavenger hunt can of course be adapted to incorporate other apps or technologies. However, one thing that we would advise against is using technology just for technology's sake! The technology is there to serve a purpose: to facilitate the activity or to streamline a behind-the-scenes work-flow. Don't be seduced by the bells and whistles that some apps (especially paid ones) may offer: *It posts to Facebook! It's GPS enabled! It tweets your score!* Keep your project objectives—and the functional requirements that result from them—at the forefront. We chose to work with popular free apps because they did what we needed them to do. They were ubiquitous enough that we felt confident about their longevity, and they were the right price. However, we knew that we might need to adapt to changes these companies made to their free products, as we wouldn't have the control or leverage that comes with paying a developer.

Starting with a small, lightweight pilot worked really well for this project, and we recommend that you do the same if you try to make your own mobile scavenger hunt. Quickly assemble a prototype; test it with a limited audience to see how it works; and adjust and revise from there. Allow sufficient time (at least a month) for testing and troubleshooting between a pilot phase and a large project launch. The logistics of the scavenger hunt itself depend on many interconnecting pieces—the library website, the location of resources in physical library spaces, the public services staff—so full run-throughs of the activity test these dependencies and reveal issues while the stakes are still low.

Finally, it helps to consider—and limit—the target audience for your scavenger hunt project. Starting with a small audience, then expanding outward, can be more manageable than working with a large group and being overwhelmed by demand. We highly recommend partnering with teaching faculty and not running the scavenger hunt as an opt-in activity. Working with instructors will guarantee good attendance and ensure a predictable number of participants. Gateway and freshman courses, such as first-year English, are logical candidates for partnerships.

LOOKING FORWARD

As the scavenger hunt project matures into a regular program, it is important to schedule periodic "check-in" time, particularly before the beginning of a new semester, to perform maintenance tasks and see if there are any desired modifications. These may include the following:

- Installing app or iOS updates on the iPods
- Checking and modifying questions that may be dependent on shifting elements (e.g., online journal access, seasonal/temporary exhibitions in library spaces, policy changes)
- Modifying or adding questions to highlight new or important library resources or initiatives
- Training new students or staff
- Refreshing project documentation

There may also be opportunities to make processes more efficient and to find more ways to make the program more sustainable. Some of the changes we have made include identifying and training student workers and paraprofessional staff to assist with scavenger hunts, streamlining scheduling procedures, and creating checklists for posthunt iPod resetting.

While we are thrilled that the mobile scavenger hunt project has been so successful, its popularity has led to increased demand, especially since it has been more widely publicized. After the project won the 2012 Association of College and Research Libraries' CLS ProQuest Innovation in College Librarianship Award, we began to receive requests for scavenger hunts from outside the First Year Writing Program. This has forced us to consider the scalability of the project and the suitability of the activity in different contexts. We will continue to facilitate the mobile scavenger hunt for our primary audience of ENG 101 courses. When faculty from other disciplines have requested them, we have let the relevant subject specialists take the lead on adapting the content and have consulted on training and logistical issues. We have also begun investigating ways to deliver an asynchronous, self-guided version of the scavenger hunt to provide options for "opt-in" participants that fall outside of our ENG 101 target audience.

A positive and rewarding by-product of attention the project has gotten is that it is being reproduced in other academic libraries. After presenting on the scavenger hunt at various library and education conferences (North Carolina State University Libraries, 2012), we have received inquiries from librarians all across the country who are interested in creating mobile scavenger hunts of their own, and a number of pilot projects have since emerged. We hope to see the effective implementation of the mobile scavenger hunt project at libraries with diverse staff sizes, resources, and student populations. We also hope that the success of this project will inspire librarians and educators to create innovative and engaging services for their communities by exploiting the relative user-friendliness of contemporary mobile devices, free apps, and web technologies.

REFERENCES

New York Public Library. (2012). *Find the future: The game.* Retrieved from http:// exhibitions.nypl.org/100/digital_fun/play_the_game

North Carolina State University Libraries. (2012). *NCSU libraries mobile scavenger hunt.* Retrieved from http://www.lib.ncsu.edu/ris/projects/scavenger.php

Tebowing. (2012). *Tebowing.* Retrieved from http://tebowing.com

Wikipedia. (2012). *Planking (fad).* Retrieved from http://en.wikipedia.org/wiki/ Planking_(fad)

7

Responsive Web Design for Libraries: Beyond the Mobile Web

Matthew Reidsma
Grand Valley State University Libraries

At some point between the moment I wrote these words and the moment you are reading them on a printed page, this chapter was coiffed and primped by a designer for optimum readability. That designer, working in the medium of print, knows with utmost certainty the size of the book—the "canvas"—and makes design decisions accordingly.

If you were reading this on the web, I wouldn't be able to predict the size of the page you were viewing, because web-enabled devices have never looked so different. We can no longer count on one or two common screen resolutions when we design a website. We're on track to have more mobile devices than people on the planet (World Bank, 2012), and in the United States, nearly half of adults who own cell phones have a smartphone (Pew Internet & American Life Project, 2012). What's more, recent data are showing that the number of adults who own more than one web-enabled device is growing, with over half of computer owners also owning a smartphone and 13% of Americans owning a laptop or desktop computer, a tablet, and a smartphone (Mitchell, Rosenstiel, & Christian, 2012). We have more devices with a greater variety of screens to design for than ever before.

GIVING UP CONTROL

Web designers are just waking up to the realization that they no longer have the kind of control that a print designer enjoys. But true to our desire for control, we address this problem by creating more of the same fixed-width layouts but, this time, one roughly phone shaped, maybe a tablet-ish one, and finally, the "regular" desktop site. We check for known mobile devices and browsers and send these visitors to separate websites. Our feeling of control returns, until new phones and tablets hit the streets, requiring us to look for another dozen or two dozen devices to make sure that each device gets to its appropriate site.

For the past 20 years, we have built fixed-width websites for a handful of screen sizes dictated by screen manufacturers. In the mid-1990s, we made sites that were optimized for screens 480 × 640 pixels, and a few years later, we expanded to 600 × 800 pixels. In the last decade, we hovered comfortably designing for screens 1,024 pixels wide and 768 pixels high, and we used every inch of that canvas because we knew the sizes of the screens our visitors had on their computers.

But the web itself has never had the kind of fixed canvas of the printed page. The screen never was the canvas. As Ethan Marcotte (2011) has pointed out, screen size is "one step removed from our actual canvas: the browser window" (p. 3). We've never had a fixed canvas, because the users of our websites always had control of the size and shape of the browser. We've never actually had control.

Accepting this doesn't mean that we have to abandon building beautiful sites and go back to pages of unstyled text. Long before mobile devices brought about the current crisis in web design, John Allsopp (2000) pleaded with web designers to "embrace the fact that the web doesn't have the same constraints [as print], and design for this flexibility." The ability to build fluid websites has always been a part of the web, but now, more than a decade after Allsopp's call to arms, we have a more sophisticated arsenal of standards-based tools to make websites that adapt to changes in the user's world.

TACKLING MOBILE IN THE LIBRARY

When I came to Grand Valley State University in late 2010, it was already trying to find a solution to the increased mobile usage of its website. That year, its library website traffic from mobile devices was a only 0.5% of all visits, while the university as a whole saw 4.5% of its traffic from mobile devices. Now just 18 months later, the library's mobile traffic has seen a tenfold increase. Visits from mobile devices for the first 6 months of this year hit 4.95%. That's nearly 1 in every 20 visitors to our website.

As an academic library, we know this growth will only continue. Recent studies report that a whopping 77% of teens aged 12 to 17 have a cell phone and 25% of those are smartphones (Lenhart, 2012). In addition, 45% of 18- to 29-year-olds who browse the web on their phone use the phone as their primary device (Smith, 2012). We needed to do something about our fixed-width website, which was barely usable on small-screen devices.

I developed a few core goals for the new site. First, all content and services would be available in full on all devices. I didn't want to make assumptions about what users on mobile devices wouldn't use. Our library would never think of limiting access to a patron in a wheelchair because we assumed he or she wouldn't be interested in a service, yet it is common practice to assume that mobile users only want hours and directions and to remove the rest of the website's functionality from a mobile site. I also wanted to avoid having two URLs for each piece of content—one for the mobile site and one for the desktop site—which can be frustrating for multiple-device users. We would have one site, not separate mobile and desktop websites.

I also wanted to take advantage of the capabilities of newer devices without shutting out older devices and assistive technologies. So I would build the site with clean, standards-based HTML and then enhance the site with CSS and JavaScript on more capable devices as much as possible. Like most libraries, we rely on many hosted vendor products to provide services to our patrons, so I wasn't in charge of much of the code being served to our patrons. Even though each vendor offered different levels of customization possibilities, I wanted all our disparate sites to appear to be a single, cohesive website. With no budget. And a team of one (me).

RESPONSIVE WEB DESIGN

In early 2010, Ethan Marcotte, a web designer from Boston, coined the phrase "responsive web design" to refer to a suite of techniques for building fluid, standards-based websites that adapt to user devices. By designing sites with a fluid grid, using CSS media queries to change styles according to device screen sizes, and using flexible images, Marcotte showed that we no longer needed to serve up separate websites to mobile devices. We could have a single code base, a single URL for each resource, and a great-looking site, no matter the device. This was the approach I took to redesigning our site. You can follow along on your own device at http://gvsu.com/library.

MOBILE FIRST

In the past, I started all of my web design projects in Photoshop or the browser, putting together mock-ups of a site that would be a fixed desktop width. But I've started thinking differently about how I structure and build websites. Rather than starting with a vision of the site on a large screen and then stripping away features or styles as the screen size drops, I now start with the smallest screen as my base and then add styles to change the layout for devices with wider screens.

And this doesn't just affect how I write CSS. The order that your HTML is presented to the user is important. In most sites, we are presented with a list of navigation items to other pages before we ever get to the content. Yet no one comes to your website for the navigation: they come for the content.

When we moved from table-based layouts to CSS, we thought that by simply keeping our style attributes separate from our HTML, we were separating style from structure. But we continued to put the navigation at the top of our document structure since that's where we expected it to appear on the page. By moving navigation below our content, we can improve the experience that less-capable and assistive devices have in visiting our sites and use CSS to progressively enhance the site for devices that have CSS-capable browsers. In fact, Marcotte has recently been proposing that web designers think of layout itself as an enhancement (Wroblewski, 2012).

This project was completed in a little over a week because I had created a user interface pattern library, a well-documented CSS library that gives me a head start on the common design patterns we use throughout our web projects. By including these styles, I no longer had to worry about styling typography or lists. You can see our reference document or grab the code for our pattern library at http://gvsu.edu/library/ui. As I walk you through how responsive web design helped us solve our content layout issues on multiple screen sizes, I'll ignore typographical styles and focus on the styles for layout.

Since we're starting with small-screen styles, we can treat each chunk of content as a block element. Because the screen is so narrow, we don't need to do much to adapt the layout of our content beyond setting up an appropriate HTML source order. However, navigation posed some special challenges. Because we've moved the navigation below the content, we need a way to give users quick way access to it without interfering with the content of the page. The simplest solution is to add an anchor link in the markup after the header that jumps the user down to navigation, like this:

```
<div id="gvsu-library_menu"><a href="#navigation">Menu</div>
   ...
<div id="navigation">
   ...
```

Now the page is looking pretty good on small screens, but as I make the browser window wider, it starts to look less inviting. In fact, once the browser window gets beyond 800 pixels, the site starts to look silly (Figure 7.1). I needed a way to enhance the layout as the screen size increased.

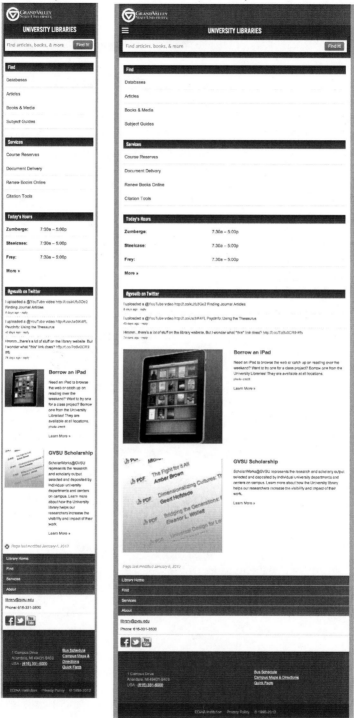

Figure 7.1. Although small-screen styles look ridiculous on larger screens, the site is still usable.

MEDIA QUERIES

If you've been writing CSS for a while, you are probably familiar with the media type attribute. It allows you to make some of your styles conditional based on the type of device the page was loading on. You could have separate styles for screens, for projectors, and for print. You could even specify a style sheet for "handheld," but support was poor in early mobile browsers.

The CSS3 specification pushed these media attributes further, creating a way to check in with devices to see if certain conditions were being met. Now, instead of serving the same CSS to all screens, for instance, we could specify a subset of CSS to be loading on screens of a certain width or aspect ratio. Here is a block of CSS that will be applied only to screens that are at least 600 pixels wide:

```
@media screen and (min-width:600px) {
    /* Awesome styles here */
}
```

This gives us a lot of flexibility on loading size-specific CSS, since we can target screen widths that will help capture a wide range of devices.

However, pixel values in media queries have some drawbacks. A better solution would be to set our queries against ems. Rather than being a fixed-width measurement, ems are a relative measure that allows for user zooming. Unless you change the default body font size in your style sheet, you can assume that 1em is 16 pixels. To convert pixels to ems, just divide the desired pixel width by the pixel equivalent of an em. In this case, instead of calling the style sheet at 600 pixels, we'll call it at 37.5 ems (600/16 = 37.5). So our query would read:

```
@media screen and (min-width: 37.5em) {
    /* Awesome styles here */
}
```

Now we have a mechanism to load styles based on particular widths, but we still need to figure out how these elements are going to be displayed on larger screens. For that, we need a grid.

FLUID GRIDS

Grids can, to some extent, give us some of the control that we loved about table-based layouts, where items were placed logically in columns and rows. Grids are, after all, systems "for ordering graphical elements of text and im-

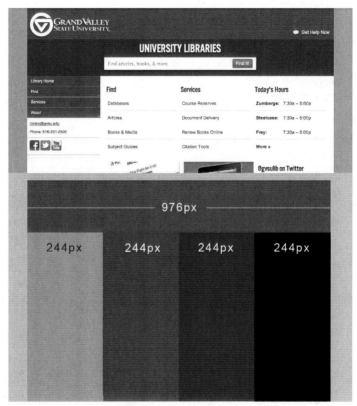

Figure 7.2. Our fixed-width mockup has four equal columns. We can make this flexible with the magic of math.

ages" (Boulton, 2005). Unlike tables, which reproduce inflexible grids in markup, our grids will simply guide us as we place our content on the page.

It is easier to explain grids by working from large-screen styles down. I developed a mockup of our site on wide screens. Here, the content of the site is wrapped in a container 976 pixels wide, divided into four equal columns. So by dividing our total page width by four, we can begin to build our grid: 976/4 = 244.

Each of our columns will be 244 pixels wide (Figure 7.2). But remember that our navigation comes in the HTML source after all of the content, meaning that we'll have to think of the content of the site in terms of three columns instead of four, while the navigation will serve as the stand-in fourth column. Here is the markup:

```
<div id="wrapper">
  <div class="line"><!-- First row of content -->
```

```
    <div class="column">
      <h3>Find</h3>
      ...
      </div>
    <div class="column">
      <h3>Services</h3>
      ...
      </div>
    <div class="column">
      <h3>Today's Hours</h3>
      ...
      </div>
  </div><!-- End .line -->
  <div class="line"><!-- Second row of content -->
    ...
  </div>
  ...
  <div id="navigation">
    ...
  </div><!-- End #navigation -->
</div><!-- End #wrapper -->

#wrapper {
  margin: 0 auto;
  width: 976px; }
.line {
  float: right;
  width: 732px; }
.column {
  float: left;
  width: 244px; }
#navigation {
  float: right;
  width: 244px; }
```

This works well for our site on large screens, such as desktops or laptops, but the fixed-width layout gives us some problems on smaller screens or browser windows. But we don't have to give up our grid to make the page respond more gracefully to the widths of our users' screens. Instead, we can convert our pixel-based layouts to fluid, relative values using percentages. If we divide the desired width of our element by the width of the containing element, we can get the relative width of our elements expressed as a percentage. So for the cited markup, when we divide the width of the column by the width of its container (.line) and multiply by 100, we get the relative width of each column: $(244/732) \times 100 = 33.333333333333\%$. So we can change our CSS to

```
.column {
    float: left;
    width: 33.333333333333%; }
```

The navigation column, however, is a child of the #wrapper div, so we need a different value for the relative width of this column: $(244/976) \times 100 = 25\%$. So we can change our CSS to

```
#navigation {
    width: 25%; }
```

We can calculate the relative width of the .line container in the same way, by dividing its value in relation to the parent element: $(732/976) \times 100 = 75\%$. So we can change our CSS to

```
.line {
    float: right;
    width: 75%; }
```

But we are still left with a fixed container, so our relative widths don't adjust to changes in screen size, since their container is a fixed width. But determining the relative width of the outermost containing element is tricky, because we don't know what the width of the element containing this element is. Nonetheless, we can use some guesswork and trial and error to find an appropriate value.

One way to do this is to look at our existing analytics and determine what different screen sizes commonly visit your site. We're not going to necessarily lock ourselves into making a site that targets one specific width, but we can use this information to help understand what relative size your outermost containing element should be. For instance, if 50% of your large-screen visits come from screens 1,024 pixels wide, and the other half come from screens 1,200 pixels wide, we can use the aforementioned formula to find the percentages that your .line div would be for each of those containers: $(976/1,024) \times 100 = 95.3125\%$ and $(976/1,200) \times 100 = 81.333333333333\%$.

Since, in this hypothetical case, our audience is split evenly between these two options, we could take the average of the two values and make our .line 88.3229165%, but we might also use this as a starting place to try out a few resolutions and see what works best. In this case, 88% works pretty well on both screen sizes, so we can change our CSS now to read:

```
#wrapper {
    margin: 0 auto;
    width: 88%; }
```

Now our four-column layout is flexible, adjusting to differences in screen sizes, but we left out something important in calculating layouts: margins and padding. When the relative widths of elements were calculated, determining the containing element was straightforward. But margins and padding are a different beast. For margins, divide the desired width of the margin by the width of the element's container. If we wanted each column to have a 6-pixel left margin, we would divide the margin width by the expected width of the .line, which is the containing element: $(6/732) \times 100 = 0.819672131148\%$. (There is no need to round these results, since computers are more comfortable than we are with complex numbers.)

For padding, we need to divide the desired padding width of the element itself, not its container. For instance, if we want to add 6 pixels of horizontal padding to each column in our layout, we would divide the desired pixel width by the width of .column: $(6/244) \times 100 = 2.459016393443\%$. But because the padding is added to the width of the element, we need to remove the same relative amount from our width element to accommodate the padding. So our CSS would now look like this:

```
.column {
    float: left;
    margin-left: 0.819672131148%;
    padding: 0 2.459016393443%;
    width: 22.540983606557%; }
```

Now we have a nice, flexible four-column layout that looks great on wider screens. But even though we've built a flexible grid under our design, things start to look cramped at smaller screen sizes. We need to use media queries to load styles for progressively larger screens.

Because most websites are designed with wide, fixed-width layouts, most mobile devices scale these sites to be displayed on smaller screens. If you've ever loaded a website on your phone or tablet and then found yourself frantically pinching and zooming to make the text readable, you're familiar with this behavior. Because we don't want these devices to turn our newly responsive website into a pinch-and-zoom mess, we need a way to tell them to display our site at the scale and width of the device. All we need to do is add the following meta tag to the head of our document.

```
<meta name="viewport" content="initial-scale=1.0, width=device-width" />
```

GETTING RESPONSIVE

Now at our smallest size, the site is a single column, while at the largest size, it will have four columns. The best way to determine where to put in some media queries to enhance the layout is to slowly make the window bigger and add new styles when things start to look crummy.

As we start to scale up, the white space after the list items becomes excessive at around 480 pixels (30em). So we'll add a media query to switch to a two-column layout here to make better use of the screen width (Figure 7.3):

```
@media screen and (min-width: 30em) {
    .column {
        float: left;
        padding: 0 1%;
        width: 48%; }
}
```

The rest of the page layout looks fine, so we'll continue to scale the width of the browser window up. At around 650 pixels wide (40.625em), the extra room in each column starts to feel excessive, so we'll move to a three-column layout. The third column will be our left-side navigation. We need to set a relative width for the .line divs that contain our columns as well as a width for the #navigation div. In addition, we should hide the anchor link for small screens that gives users easy access to navigation. Also, the search bar is getting a little bit long, so we'll scale it down and center it (Figure 7.4). The new media query looks like this:

```
@media screen and (min-width: 40.625em) {
    .line {
        float: right;
        width: 66.66666666%; }

    #navigation {
        width: 33.3333333%; }
```

Figure 7.3.   As the screen gets larger, we have more room for columns.

```
#gvsu-library_menu {
    display: none; }

#search-box {
    margin: 0 auto;
    width: 48%; }
}
```

Figure 7.4. Once the screen gets wide enough, we can move our side navigation into place.

Now the site is really coming along, but as the screen gets wider, it again feels more stretched. By 900 pixels (56.25em), the three-column layout is feeling a bit stretched, so it's time to use those styles we created earlier for our four-column grid. Since our campus home page sets a fixed width of 976 pixels on the content container once the screen is at least 1,024 pixels, I added the following style to make the #wrapper stop scaling.

```
@media screen and (min-width: 64em) {
    #wrapper {
        width: 61em; /* 976px / 16px = 61em */ }
}
```

Now we have a basic responsive website that adapts its layout to the size of the screen. And since we began with a foundation of clean, semantic HTML, we can be sure that the site will work great on any web-capable device.

INTERNET EXPLORER—OR, WHAT ELSE IS NEW?

While media queries enjoy great support in modern browsers, Internet Explorer 8 and earlier ignore them. If we start with mobile styles and then use min-width queries to enhance for larger screens, Internet Explorer will simply load the styles for small screens and ignore the rest. If you're an academic librarian like me, you have a campus full of computers running ancient versions of Internet Explorer, so simply leaving this unaddressed isn't an option. But there are a few ways to work around this limitation.

First, there are several JavaScript polyfills that will patch browsers that don't support media queries. Scott Jehl's Respond.js is the most lightweight (http://github.com/scottjehl/Respond). If you can't use JavaScript, you could restructure your media queries to begin with your wide-screen styles and then use max-width: media queries to scale the site down to smaller screens. Internet Explorer will ignore all of the styles in media queries, so it will load only the initial styles for wide screens. This is the approach that our campus uses, but it makes me uncomfortable. Less capable devices that support CSS but not media queries will get served a "desktop" website instead of a basic, small-screen optimized site. In my experience, scaling down from a desktop site also makes for more CSS.

The solution I chose was to create an Internet Explorer–specific style sheet that loads after your media queries, passing all of the wide-screen styles to early versions of Internet Explorer. That would look something like this:

```
<!--[if lt IE 9]>
  <link rel="stylesheet" type="text/css" href="ie.css" />
<![endif]-->
```

Since respond.js conflicted with some aspects of our campus content management system, this was the solution I took for the library website. Internet Explorer users still won't get a responsive site, but users will less-capable mobile devices will get a site styled for mobile.

FLEXIBLE IMAGES

The last element of responsive web design is flexible images. Since our website (and most library websites) are not image heavy, making our images adapt to their containers is fairly easy. We just need one line to our CSS:

```
img {
    max-width: 100%; }
```

Now our images will never be larger than their containing elements. As the containers scale down, so will the images. Of course, if you are building a digital library with a lot of large images, you probably will need a more sophisticated approach to dealing with images. There are several solutions under discussion for how to serve up different images to devices based on screen sizes, but none of them have been implemented by browser makers yet. If you need to serve up a lot of images, I recommend looking at Scott Jehl's Picturefill (https://github.com/scottjehl/picturefill), which replicates one of the proposed solutions in JavaScript.

USER RESPONSE

Since we've only just launched the responsive design on the pages in our content management system, it is too soon to get any useful data on how our users are making use of the site. We also haven't advertised the change and don't intend to. Our primary audience is students, and as more and more of them use mobile devices, their expectations for the sites they visit increase as well. We're expected to have a site that works as well on a mobile device as it does on a desktop computer, and no one congratulates you when you meet their expectations.

FURTHER CHALLENGES

Now that our library home page and content management system are responsive, I'm faced with the task of replicating this design in all of our hosted vendor systems. Some vendors give us a lot of flexibility in how we design the user interface, such as Illiad, our interlibrary loan system. Others give us less control. To gain control of the styles of our 360 Link link resolver from Serials Solutions, I wrote a script that scrapes the page and then rewrites it on the fly in our library's template. Others, such as Serials Solutions' Summon Discovery Service, give us no control over the design of the page. On those systems that do give us the ability to change things, however, we want to be consistent.

This is perhaps the greatest challenge for libraries employing responsive web design. Because we likely will never have a consistent template that works across all of our vendor products, I may end up writing different CSS for every product to achieve the illusion of consistency. In addition, some of our services, such as LibGuides and Serials Solutions' Summon, already redirect mobile users to a separate site. For everything else, responsive web design has allowed us to offer all of our services to our users regardless of the device they use.

REFERENCES

Allsopp, J. (2000). *The dao of web design*. Retrieved from http://www.alistapart.com/articles/dao/

Boulton, M. (2005). *Five simple steps to designing grid systems—Preface*. Retrieved from http://www.markboulton.co.uk/journal/comments/five-simple-steps-to-designing-grid-systems-preface

Lenhart, A. (2012). *Teens, smartphones and texting*. Retrieved from http://www.pewinternet.org/Reports/2012/Teens-and-smartphones/Summary-of-findings.aspx

Marcotte, E. (2010). *Responsive web design*. Retrieved from http://www.alistapart.com/articles/responsive-web-design/

Marcotte, E. (2011). *Responsive web design*. New York: A Book Apart.

Mitchell, A., Rosenstiel, T., & Christian, L. (2012). *Mobile devices and news consumption: Some good signs for journalism*. Retrieved from http://stateofthemedia.org/2012/mobile-devices-and-news-consumption-some-good-signs-for-journalism/

Pew Internet & American Life Project. (2012). *A closer look at gadget ownership*. Retrieved from http://pewinternet.org/Infographics/2012/A-Closer-Look-at-Gadget-Ownership.aspx

Smith, A. (2012). *Cell Internet use 2012*. Retrieved from http://www.pewinternet.org/Reports/2012/Cell-Internet-Use-2012.aspx

World Bank. (2012). *Information and communications for development 2012: Maximizing mobile*. Retrieved from http://www.worldbank.org/ict/IC4D2012

Wroblewski, L. (2012). *An event apart: Rolling up our responsive sleeves*. Retrieved from http://www.lukew.com/ff/entry.asp?1494

8

Using iPads to Revitalize Traditional Library Tours

AMANDA BINDER, SARAH SAGMOEN,
NATALIE TAGGE, AND NANCY J. WEICHERT
Brookens Library

During the summer of 2010, Brookens Library at the University of Illinois Springfield made the decision to purchase 10 iPads. The intention was to provide the opportunity for students, staff, and faculty to check out iPads for in-library use. Excitement about the vast opportunities provided by the new tool and its use in academic libraries was high, and Brookens hoped to be among the first to offer an iPad-lending service to its patrons. With the delivery of the iPads to the library came the first wave of concern that libraries with iPad-lending services would be in violation of the Apple licensing agreement. The Apple agreement states, "You may not rent, lease, lend, sell, redistribute, or sublicense the iPad Software." While there was much debate on how the agreement should be interpreted, Brookens Library decided not to take the risk and stopped all plans to circulate the iPads as originally intended. With this decision, the library was forced to reenvision how the iPads would be used. With a total price tag of over $5,000 and the understanding that a first-generation gadget from Apple would be out of date in a year's time, it was imperative that a new plan be put into place quickly. Integration into a preexisting Library Instructional Services Program collaboration seemed the easiest place to initially deploy the iPads.

PILOTING IPADS IN LIBRARY INSTRUCTION

The Library Instructional Services Program has had a long history of course-integrated instruction in a foundational class required of all chemistry and biology majors, CHE/BIO 301. Since the late 1990s, librarians have routinely provided three 50-minute course sessions focusing on topics such as basic database search strategies, evaluation of sources, and use of SciFinder Scholar, while corresponding in-class and out-of-class assignments reinforce information literacy skills introduced by the librarian and course instructor.

The initial phase of this course-integrated instruction traditionally involved devoting an entire course period to a tour of the physical and virtual library (25 minutes on a building tour and 25 minutes on a "tour" of the library website), followed by two concepts-focused sessions. After two semesters of dragging half-asleep CHE/BIO 301 students through the library (the course often meets at 8:00 AM) and questioning the pedagogical logic of devoting course time to this activity, in fall 2010 the instructional librarian and liaison to the chemistry and biology departments, Natalie Tagge, began to explore engaging students with active learning components during the tour. After it was determined that the iPads could not be loaned due to Apple's licensing agreement, Tagge decided to pilot their use in the tour component of CHE/BIO 301. She hoped that the iPads could enliven the tour while encouraging students to learn more actively about the physical and virtual aspects of the library.

Tagge developed and incorporated an iPad scavenger hunt centered on book and journal discovery into the CHE/BIO 301 tour. Book and journal titles were selected to expose students to multiple formats, item statuses, and locations. She also contacted the course instructor to explain the addition of technology into the tour and her learning objectives for incorporating the iPads. To ensure that there would not be any issues with the connection of the iPads to wireless or with the flow of the lesson plan, Tagge took herself on trial runs of the scavenger hunt. A fellow instructional librarian and resident technology guru, Sarah Sagmoen, also kindly agreed to make sure that all the iPads were functional and fully charged prior to the class sessions.

Upon arrival, students were placed in pairs (due to the number of iPads) and given a slip of paper with a book and a journal title on it and a worksheet that asked them to record the title, call number, and location of each item. If the item was owned by the library in print, the student was also asked to locate it on the shelf. Around 50% of the revised 50-minute tour was devoted to a traditional session guiding students through the physical building. At regular intervals during this period, students were asked to use the iPads to independently complete steps in the scavenger hunt. Tagge and the course instructor were available at all times if students needed assistance with the scavenger hunt or with operation of the iPad. The students were then quizzed on the results of their searches, which typically led to productive group discussions on everything from the finer points of Library of Congress call numbers to serial formats and conference proceedings—all relevant information for chemistry and biology majors. iPads were also utilized during the tour to direct students to pertinent information on the library website, such as librarian contact information (Figure 8.1).

Figure 8.1.   Using the iPads, students locate materials in the library stacks.

In Tagge's experience, the addition of iPads very much enlivened the crack-of-dawn tours for all CHE/BIO 301 stakeholders. iPads provided an excellent pedagogical tool to explain abstract principles (e.g., source type) and concrete principles (e.g., call number location) in the authentic context of navigating the physical and virtual library. Students were quite enthusiastic about the iPads, evinced via comments during the session and on subsequent session evaluations. The success of this pilot led to broader and continued use by Tagge and other librarians in the CHE/BIO 301 library tour and to expansion of similar use in other course-integrated and non-course-integrated instruction.

Integrating iPads into Drop-In Library Instruction Series

At the time that Tagge was creating and implementing the iPad scavenger hunt into the CHE/BIO 301 courses, the library was offering 1-hour drop-in tours and workshops. The drop-in series was developed in spring 2010 by Tagge and instructional librarian Amanda Binder, in response to the results of a 2-year grant-funded ethnographic research study of the library and the campus. Realizing that information literacy instruction was not being fully integrated into the first-year experience and that the library was not reaching a large percentage of the growing online, transfer, and commuter student populations, the librarians created the drop-in library series. The hope for the series was to further the Brookens Library mission.

The series includes 1-hour sessions that are offered both online and on campus and at various times of day and night, to accommodate the entire student population at the University of Illinois Springfield, including traditional, online, and commuter students; new, transfer and existing undergraduates and graduate students; and even faculty. Sessions are taught by librarians in the Library Instructional Services Program, one of whom serves as the coordinator of the series. The series coordinator is responsible for determining the best dates and times for the sessions, working with the instructional librarians to staff each session and coordinating all communication and marketing efforts.

The series was developed as a drop-in to create a more relaxed and welcoming feel to the library, for both students and faculty. The philosophy behind the series is that students can come on their own time, outside of class, to learn about the library resources available to them and to prepare for their research assignments. While the series would appear to be designed for the self-motivated learner, it is marketed to faculty in such a way that they can easily see the benefits for themselves and their students of offering credit for attending sessions, thereby providing an additional incentive.

iPad Integration

After conducting the iPad tour for CHE/BIO 301 students, Sagmoen realized the opportunity to integrate the active learning components into the drop-in tour already being offered by the library. Where Tagge had designed elements around the needs of CHE/BIO 301 students, Sagmoen worked to generalize the activities to speak to students from various departments.

One of these changes was the creation of two lists of book titles and one list of article keywords. Sagmoen originally created one list of book titles, but when the books were not reshelved between back-to-back tours, it was clear that a second list was needed if two tours were to take place on one day. Additionally, Sagmoen included books from the oversized collection as well as the e-book collection on this list to ensure that students were shown the importance of looking at the book's location in the item record.

The drop-in iPad tours start in the library classroom with a quick overview of library resources and services. Then the librarian demonstrates how to use the library's single search interface to search for books first and articles second. Once the classroom portion has been completed, participants are provided with an iPad, and the librarian quickly shows students how to use the technology. As iPads have become more ubiquitous, this section is often shortened or changed to match the skills of the students at individual tours. The librarian then begins the tour, and much like the original iPad tour created by Tagge, it is divided evenly between a traditional tour highlighting the physical space and one placing active learning tasks throughout (Figure 8.2).

Figure 8.2. Sagmoen and Weichert lead a group of students through the library while conducting an iPad tour.

While touring the general collection of the library, the students are each given a slip of paper with a title to search in the single search interface using the iPads. Students use the iPads to gather the information they need from the catalog record to go into the stacks, locate the book, and bring it back to the librarian. The librarian uses this time to walk through the stacks and help students as needed. The librarian then talks about the next steps after locating a book. If the students want to check out the book, they are informed that they need their student ID. At this point, the librarian also talks about the length of book loans and how to renew items. For the purpose of this activity, it is determined that the student does not check out the book, and in that case, students are instructed to place the book in the appropriate boxes for reshelving. Students then move to the library's print periodical collection. Again, students are provided with a slip of paper, this time with a broad keyword, such as *homelessness* or *global warming*. Using the same single search interface, the students search for an article on the provided topic. Students are instructed to select the first article in the results list and determine if it is available from the library; if so, they then walk through the steps to locate the full text online or the information needed to locate the article in print. Again, the librarian will take advantage of the teaching moment and discuss what happens when an article isn't available from the library, explaining that a student's next step is to place an interlibrary loan request. Finally, the remainder of the traditional tour is completed, and the students return the iPads.

MARKETING AND PUBLICITY

One of the essential components to launching the drop-in series and iPad tours was a strong marketing campaign. The multimedia communications specialist for the library was tasked with creating a fun and innovative flyer that could be posted around campus and marketed through the library blog and other social media. Links to the blog are posted to campus announcements, sent to the faculty e-mail distribution list, and communicated to departments through the library liaisons.

The tours and workshops are also entered into the library's online calendar, and settings allow it to feed into the campus calendar that rotates on the university website. The library also saw value in developing a relationship with the multimedia writer/producer for campus relations, who is responsible for the campus's social media. This individual promoted the library's social media posts about the tours and workshops, creating a broader audience for the library and its offerings. One faculty member that creates Twitter accounts for particular courses also retweets the library posts. The library has also posted on the walls of the various Facebook pages created by university groups, such as Student Life and the Center for Teaching and Learning. The library eventually created a blog specifically for faculty (Faculty Focus) that is also used to promote the series. Faculty members are encouraged to promote the series in class, through their syllabi, and through Blackboard announcements. They have also been encouraged to post a copy of the flier on their office or classroom doors.

While all of these forms of outreach helped the library to create buzz surrounding the iPad library tours on campus, none was more effective at building student attendance than direct contact with the faculty. The series relieves faculty of dedicating class time and office hours to basic library instruction. The library also made it very easy for faculty to offer credit. Attendance is taken at every drop-in session. The attendance sheet includes a section for students receiving credit to note the name of their instructor. The library coordinator for the series collects the attendance sheets and then e-mails those instructors the names of students that attended. As a result of direct marketing to faculty, the majority of students that attend the drop-in series are receiving credit from their instructors.

FEEDBACK AND ASSESSMENT

Online surveys created with SurveyMonkey are used to collect feedback from students and faculty. Following each session, students receive an e-mail from the library series coordinator with a link to the survey. In the future, the library plans to have students complete the surveys on the iPads at the end of each session. A survey was also sent to the faculty in spring 2012 for feedback.

The feedback from the surveys has been invaluable in improving the sessions. Suggestions for more interaction and active learning in the sessions were some of the comments that inspired the librarians to apply the iPad tour model to the drop-in library series. Both students and faculty have responded positively to the new, interactive format of the drop-in tours. Students enjoyed having the opportunity to search for materials in the library catalog and attempt to find them on the shelves during the tour. Faculty appreciated the convenience of being able to send their students to the library for instruction, without having to coordinate it.

FURTHER IMPLEMENTATION OF IPAD TOURS

During the spring and summer of 2012, the library experienced a great deal of interest in the library iPad tours from various institutional entities. When the university approved a first-year seminar program for the 2012–2013 academic year, the librarians immediately saw an opportunity to integrate the iPad library tours into the first-year curriculum. Due to the library's successful advocacy, information literacy had previously been established as a baccalaureate goal for the General Education Program and consequently became a goal of the First-Year Seminar Program. In conjunction with the General Education Council, the Library Instructional Services Program created the following freshman seminar learning outcome: "Students will be able to locate, evaluate, organize, and use research material from electronic and print sources." While the librarians were thrilled to be included in the freshman seminar program, there was concern about the logistics involved in reaching each student. Obvious barriers were considered, such as librarian scheduling, staff shortages, classroom space, and class time constraints. The librarians agreed that with a little tweaking, the library iPad tours would be the most probable modality in which to reach each student.

The next step was to have the library dean contact the associate vice chancellor for undergraduate education, who was responsible for coordinating the first-year program. The vice chancellor immediately recognized the iPad tours (she herself had promoted them to students enrolled in her classes), and she invited Binder, coordinator of the drop-in series, to speak at a workshop designed for instructors of the first-year seminar. The library presentation proposed that all first-year seminar instructors require students to attend one of the drop-in iPad tours and assign a grade for attendance. The vice chancellor approved of this proposal during the workshop, thereby making the tours a required component in the freshman seminar courses. Next, the librarians began their work on codifying the library iPad tour process.

As mentioned, the library iPad tours were developed as drop-in sessions with no registration required. Librarians Sagmoen and Nancy Weichert were aware that this model would not be sustainable with the influx of freshman seminar students. Historically, librarians offered an average of six tours per semester, with variable attendance levels that averaged six to eight students using 10 iPads. To accommodate increased attendance, the librarians determined that 10 additional iPads were needed as well as an increased number of tours, required registration, limits on attendee registration, and a more robust participant-tracking system. Binder and Sagmoen had selected six dates for the fall 2012 semester tours. After evaluating the number of incoming freshman, Sagmoen and Weichert determined the need for an additional four sessions. It was determined that 10 sessions would be adequate to cover the needs of freshman seminar students as well as the general student body. Sagmoen and Weichert felt it important to allow for drop-in library patron participation without having to schedule a separate sequence of tours. Other variables included in the decision were the number of available iPads (20), classroom space, and faculty preference (some faculty requested tours to be held during class time).

The library has experienced a great deal of interest in the library iPad tours. Per a request from the Provost's office, librarians adapted the library iPad tours for new faculty orientation to begin summer 2012. With iPads in hand, new faculty members will be taken through the library and given instructions on how to place course reserves, contact their library liaison, and search for and acquire materials. For example, while new faculty members are at the circulation desk to meet the staff, librarians will instruct the new faculty members to place course reserves, request articles through interlibrary loan, and set up their library accounts using the iPads. Simultaneously, these tours provide a way to market library iPad tours for students, discipline-specific workshops, and the library mobile site. Similarly, the Office of New Student Orientation and Parent Relations has expressed interest in integrating the library iPad tours with the orientation process. Librarians have yet to develop this program but are eager to begin work with the new student orientation planning committee.

In an effort to maintain interest and momentum, librarians have been planning additional ways in which to integrate iPads into instruction. The library liaison to the computer science department, Jan Waterhouse, utilized the iPads in a discipline-specific drop-in session. Waterhouse had students use the Brookens Library mobile website to access the mobile versions of IEEE, ACM, and Springer. She used Poll Everywhere throughout the presentation to keep the students engaged by polling before and after each demonstrated resource. At the conclusion of the session, Waterhouse had students use the iPads to access and complete a SurveyMonkey survey in which she was able to obtain 100% participation. The survey results reflected a positive student experience. Students were especially impressed with the iPad integration. Going forward, this model has been followed by other librarians since fall 2012.

ADVICE TO OTHER INSTITUTIONS

The Challenge of Managing Multiple iPads

Soon after the iPads arrived, Sagmoen realized how cumbersome it was to manage a group of mobile devices. First she had to set up an Apple ID for the library that was not connected to a credit card per the library's fiscal officer's request. Then Sagmoen had to decide if she would connect all of the iPads to this same Apple ID or create a number of accounts. After some research, it was clear that connecting them all to the same account was the best choice.

From there, coming up with a system that kept the iPads charged and up-to-date with software updates became a tedious task and was made a bit more challenging because Sagmoen was working on a PC instead of a Mac platform. Until iOS updates were pushed out over wi-fi, Sagmoen had to plug in each iPad and run the update. Depending on the size of the update, this could take up to 2 days to complete. Because the iPads weren't used much at the beginning, they weren't consistently being charged. Therefore, in preparation of each use, they would have to be charged. Again, this typically took 2 days to complete. With only so many outlets available, charging had to be done in shifts. Because Sagmoen knew that the library would eventually use the iPads more frequently, it was clear that there needed to be a better system in place for managing them.

To help with this challenge, Sagmoen reached out to the Instructional Support and Training Team in the information technology services (ITS) department at the University of Illinois Springfield. While the library was the first department on campus to purchase iPads, ITS was right behind with a purchase of its own. Because ITS was also working on a system for managing a bank of iPads, it was happy to support the library in creating a solution. After working with ITS, the library ordered a MacBook Pro and now uses Apple Configurator to manage the iPads on the software side. The Apple Configurator allows the administrator of the iPads to create a profile for a group of iPads that dictates settings, restrictions, and preferences and then mass deploys that profile to the entire group at once. In addition, the library purchased an iPad cart that keeps the iPads secure and charged. The cart is signed out to the librarian prior to the tour with an inventory sheet so that the librarian is able to ensure that the number of iPads handed out at a tour are safely returned at the end.

Tracking Student Participation Using LibCal

Historically, librarians have worked with faculty to encourage student participation in the library iPad tours through the use of incentives such as extra credit. Librarians tracked student participation through a sign-up and reported to faculty a list of their participating students. Though not ideal, this process was manageable due to the relatively low number of participants. Going into the fall 2012 semester, Sagmoen and Weichert determined that tracking 250 freshman seminar students in this fashion would not be a sustainable model. Librarians began a trial of LibCal, an event management tool by the creators of LibGuides. After a brief trial period, it was decided to integrate LibCal to create a calendar of sessions and track participant registrations. With the familiar interface, the ability to easily create and manage events, and the interoperability with LibGuides, the adoption of LibCal by librarians was a natural next step.

In the case of freshman seminar, students are required to register for an iPad tour through LibCal. The registration form asks for the student's full name, e-mail address, and course instructor's name. Each session is capped at 24 students—a number determined to account for absences and drop-in students. Once registered, students receive a confirmation e-mail. In the event of a room change, session cancellation, or time change, all registered participants will receive an auto-generated e-mail with new dates or times. Prior to each tour, librarians will print the list of session participants to be used as an attendance sheet. Following each session, librarians will export the registrant list, directly from LibCal, and check it against actual attendee names. Finally, each professor is e-mailed a list of attendees from his or her course. Though not entirely labor free, this process should prove to be more efficient than the previous manual input method. It is important to note that fall of 2012 was the first semester that librarians utilized LibCal, with alterations and improvements expected.

Adaptability

Much of the iPad tour success can be attributed to a solid marketing campaign, faculty-supported attendance incentives, and efficient iPad management. Faculty support has been the most influential element in the success of the program. The encouragement and incentives offered by faculty to their students for participation has proved to be far superior to any other marketing or publicity efforts. Internally, the coordination and organization of the program continues to be altered and improved on regularly. Managing the iPads efficiently through the use of an iPad cart and central updating system has saved an innumerable amount of staff time and resources. The latest integration of LibCal has shown to be a promising organizational improvement to the planning, registration, and tracking process.

Along with technological modalities, the librarians have learned to be fluid in their use of iPads as instructional tools. Libraries know that many of their patrons are using mobile devices in their everyday lives, but often due to lack of knowledge, patrons are not utilizing library mobile services. Brookens instructional librarians have learned how mobile devices such as iPads can be utilized during the research and discovery process. As more library resource providers shift toward mobile interfaces, Brookens librarians will continue to seek opportunities to develop mobile programs to meet the needs of their patrons.

9

Going Mobile at Illinois: A Case Study

JOSH BISHOFF

University of Illinois Libraries, Urbana

The University of Illinois Libraries began exploring options for a mobile web presence the fall of 2009, during a period of widespread growth in the use of Internet-capable cellular phones. Current evidence continues to suggest the pervasiveness of mobile Internet devices: According to the Pew Internet and American Life Project, 46% of U.S. adults owned a smartphone as of February 2012.[1] Anecdotal evidence in 2009 also compelled us to explore mobile services: It was easy to observe library users accessing library resources with their mobile devices; students and faculty regularly approached reference desks with their mobile browsers open to OPAC or database pages, asking for assistance. Integrated library system vendors, publishers, and abstracting and indexing services were developing mobile interfaces.

Within this set of circumstances, a collection of interested librarians and academic professionals from public and technical services positions organized the library's Mobile Services Working Group, charged with exploring various vended and homegrown solutions for developing mobile access to library services and collections. One of our first and most helpful activities was to examine existing mobile service offerings from other academic libraries; there is an excellent index of such libraries available at http://www.libsuccess.org/index.php?title=M-Libraries. The functionality of these sites and applications varied widely: Some mobile sites offered only library hours, location, and contact information; other sites featured OPAC and database access, reference assistance via SMS/text message, and virtual tours.

Discussions within the working group quickly distilled the chief user objectives that the library's mobile presence should serve:

- Find libraries.
- Find library materials.
- Find help.

This chapter addresses how the group pursued these goals in the mobile context, and it discusses some of the challenges we encountered.

DEVELOPING FOR MOBILE: MOBILE APP OR MOBILE WEBSITE?

One of the most significant early discussions toward a mobile presence for the university libraries concerned whether to develop a mobile website or a device-integrated mobile application, or app. Mobile apps, at the time, were garnering significant interest from mobile phone users: Apple's App Store had debuted in the summer of 2008, and there was widespread interest in software development for Apple's iOS, Google's Android operating system, the Blackberry OS, and others.

The discussions over whether to build an app or a mobile-optimized website eventually hinged on questions of access and cost. While there was some initial enthusiasm for developing an iPhone or Android app, we reasoned that we would either need to develop equivalent services for other mobile platforms or risk neglecting that fraction of our users that did not use iPhones or Android devices. An examination of our weblogs to note which kinds of mobile devices were attempting to access our library's main web gateway revealed a variety of mobile platforms: Apple's iOS, Google's Android, the Blackberry OS, Palm, Symbian, Windows Mobile, and others. With such a variety of mobile devices trying to access library services, we felt that it would be most prudent to develop a mobile solution that would function on the largest possible number of devices. The iOS, Android, and Blackberry platforms cannot utilize the same software: To develop an app experience for these devices, we would need to write completely separate applications, identical in functionality but utilizing completely different programming languages and software paradigms. In-house app development was simply too costly, and the vendor solutions we explored for app experiences were similarly locked in to function on only a limited number of available mobile devices.

It is also worth noting that none of the library's technical staff at the time possessed significant expertise developing for any of these incipient mobile operating systems; the library's software strengths are primarily in developing dynamic websites with server-side scripting technologies, such as PHP, the Microsoft .NET framework, and others. Put simply, we knew how to make dynamic websites that offered access to library resources; we just hadn't yet developed them with a focus toward mobile devices.

DESIGN CONSIDERATIONS FOR MOBILE DEVICES

The most obvious constraint when designing for mobile devices is the size of the device screen. Most websites are designed with the desktop in mind, and many—the University of Illinois Library included—have an unfortunate habit of utilizing every available inch of screen space (Figure 9.1).

A glance at the University of Illinois Library website might quickly reveal that the process of adapting to the mobile user is sometimes less a technological problem than a problem with deciding what, exactly, are the most important features of a website. Within this discussion of what to include on the mobile website, an important point merged: We did not believe

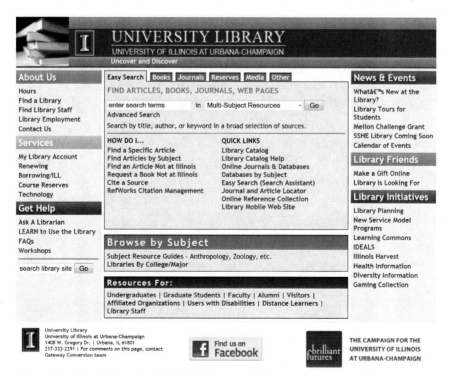

Figure 9.1. The University of Illinois Library website.

that the mobile website needed to offer identical services to the desktop site to be successful. We came to view the mobile website as a complement to existing library services, and we resolved that the development process and user feedback could further inform us of which services mobile users found the most valuable.

With this in mind, the library's Mobile Services Working Group recalled its fundamental user interaction goals—to assist finding libraries, material, and help—and weighed to what extent each of the dozens of links on the library's desktop-sized site contributed to these functions, eliminating those that contributed the least. The group produced the following short list of links:

- Search the Catalog
- Find a Library
- Mobile Databases
- Ask a Librarian

We did not spend a great deal of time searching for a visual design pattern that would accommodate these headings; our examinations of other mobile sites had revealed essentially two kinds of visual interface: a one-dimensional list of options that stretched across the device screen or a two-dimensional matrix of icons (Figure 9.2).

We chose the first option—the one-dimensional list—because it was simplest to code; moreover, we'd noticed that complex layouts rendered unpredictably on available handsets. This decision led to the prototype in Figure 9.3, which satisfied our goals of being visually and conceptually straightforward.

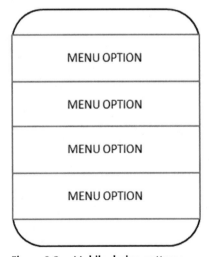 

Figure 9.2. **Mobile design patterns.**

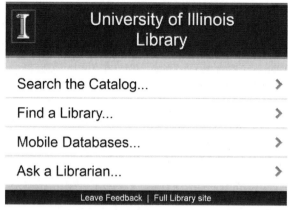

Figure 9.3.   First prototype.

SHRINKING THE LIBRARY CATALOG

The University of Illinois Libraries use two separate online public access catalog (OPACs), both administered by the I-Share consortium: Ex Libris's Voyager system and an implementation of the open-source VuFind OPAC. Both presented problems when we explored vendor solutions to mobile access. Neither Ex Libris nor VuFind had mobile interfaces, and the vendor solutions we explored presented all of the problems of administering an additional OPAC: regular MARC (machine-readable cataloging) dumps to populate a separate database, difficulty accessing real-time holdings information, and inability to offer access to the full I-Share consortial OPAC.

A visiting librarian and software developer on our working group had, related to a research interest, developed an XML gateway to the VuFind OPAC using ASP.NET technology and volunteered to explore the possibility of utilizing this work in the delivery of catalog results to a mobile interface. This was a felicitous development in the search for a mobile service that could provide OPAC access; it ended up as a step toward a satisfactory mobile interface to the VuFind OPAC. The XML gateway that the librarian developed broadcast a user query to VuFind, retrieved the response in full HTML, and translated it with regular expressions into a structured XML that could be restyled to be suitable for a mobile display. The OPAC-to-XML service is a RESTful web service that returns a stripped-down version of the catalog: Once again, the mobile use case forced us to make decisions about what information we could reasonably represent on a small screen. We reasoned that the most salient features of a catalog record were the title, author, creation date, and format information. This was a difficult compromise and obviously imperfect, but it was necessary

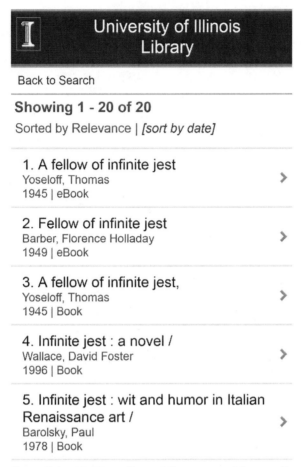

Figure 9.4. Mobile online public access catalog results display.

to effectively represent a results list on a mobile device screen. The next step was to write an XSL style sheet that would transform the XML catalog records into a useful view for the mobile device. We tried to keep this display visually consistent with our initial front-page prototype (Figure 9.4).

FIND THE LIBRARY: DELIVERING HOURS AND LOCATION TO MOBILE DEVICES

We anticipated that one of the principal tasks of a mobile website user would be to determine if a particular campus library is open and how to get to there. This seems like it should be a simple task, but in 2009 the Illinois libraries comprised over 30 departmental libraries spread across almost a dozen campus buildings. On the full-sized library website, a user might discover hours information, as it is displayed in a large table presenting the hours of each department (Figure 9.5). This table is regrettably unsuited for most mobile displays; it's a good example of the limitations of a presentational container—in this case, the HTML table—rendering inaccessible the information it contains.

Our solution to this was to develop a small database of metadata about each library. We entered the following pieces of information about each department library into a Microsoft Access database:

- latitude and longitude of the building,
- opening and closing hours for each day of the week,
- telephone number, and
- the URL of its full-sized webpage.

With this information, the mobile website could guide a user to a particular library and its services in an accessible manner.

Including latitude and longitude of campus libraries allows the site to generate dynamic maps and links that capable handsets can utilize for navigation assistance. Each library's entry in the database results in a version of Figure 9.6.

The map displayed to the mobile user is dynamically generated by a call to the Google Maps application programming interface[2] using the latitude and longitude information from the database: http://maps.google.com/maps/api/staticmap?center=[*latitude,longitude*]&zoom=16&size=350x400&sensor=false&markers=[*latitude,longitude*]&key=your_google_api_key. This approach can further leverage the capabilities of Google Maps and the user's device hardware when one uses the dynamically generated map image as an HTML anchor and links the image to a simple Google Map URL with the same position information: http://maps.google.com/maps?q=[*latitude,longitude*].

Advanced handsets that can recognize links with embedded position information—such as the iPhone and various navigation-equipped Android devices—will, if the user taps the map, offer to guide the user to that library using the device's built-in navigation features.

ADDITIONAL FEATURES

The library has a Google Voice account that can respond to incoming text messages; we utilized this technology to offer reference via SMS/text. Users who click the "Ask a Librarian" link are offered options of calling the

Unit Hours for Fall Semester 2012

Unit	Hours
ACES - Funk (includes CPLA Reference & Resource Center): 1101 S. Goodwin (333-2416) *	Aug. 27-30: 8:30am-midnight; Aug. 31: 8:30am-10pm; Sept 1: 10am-10pm; Sept 2: 10am-midnight; Beginning Sept 4, M-Th: 8:30am-3am; F: 8:30am-10pm, Sa: 10am-10pm; Su: 10am-3am
Applied Health Sciences	See: Social Sciences, Health, and Education Library
Architecture & Art: 208 Architecture (333-0224) *	M-Th: 8:30am-10pm ; Fri 8:30am-5pm ; Sat 1-5pm ; Sun 1-10pm
Business & Economics	For Business see: Business Information Service; For Economics see: Social Sciences, Health, and Education Library
Business Information Services: (333-3619)	By appointment only
Center for Children's Books: 24 GSLIS (244-9331) *	M: 10am-5pm; T: 10am-7pm; W: 4-7pm; Th: 10am-6pm; F: 10am-5pm
Central Access Services: 203 Library (333-8400)	M-Th: 8:30am-11pm; F: 8:30am-6pm; Sa: 11am-5pm; Su: 1-11pm
Chemistry: 170 Noyes Lab (333-3737) *	M-Th: 8:30am-9pm; F: 8:30am - 5pm; Sa: 1-5pm; Sun: 1 - 9 pm
Classics: 419A Library (333-1124)	M-Th: 9am-7pm; F: 9am-5pm; Sa & Su: 1-5pm
Communications: 122 Gregory Hall (333-2216) *	M-Th: 8:30am-11pm; F: 8:30am-5pm; Sa: 1-5pm; Su: 1-11pm
Education & Social Science	See: Social Sciences, Health, and Education Library
Engineering-Grainger: 1301 W. Springfield (333-3576) *	Aug. 27-30: 8am-midnight; Aug. 31: 8am-10pm; Sept. 1: 10am-10pm; Sept. 2: 10am-midnight; Sept. 4: 8am-midnight; Beginning Sept. 5, open Sundays at 10am and remain open 24 hours until Fridays 10pm. Saturdays: 10am to 10 pm
Health Sciences: 102 Medical Sciences (333-4893) *	M-Th: 8:30am-11pm; F: 8:30am-9pm; Sa: 9am-6pm; Su: 1-11pm
History, Philosophy, and Newspaper (includes African American Research Center): 246 Library (333-1509)	M-Th: 9am-7pm; F: 9am-5pm; Sa-Su: 1-5pm
Illinois Fire Service Institute: 11 Gerty Dr. (333-8925) *	M-F: 8am-5pm
Illinois History and Lincoln Collections: 322 Library (333-1777)	M-F: 8:30am-5pm
Illinois Researcher Information Service (IRIS): 433 Grainger Engineering Library (333-9893) *	Closed. See Grant forward go.illinois.edu/gf
Interlibrary Loan and Document Delivery: 128 Library (Lending 333-1958; Borrowing 333-0832)	M-F: 8:30am-5pm
International & Area Studies Library: 321 Library (333-1501)	M-Th: 9am-7pm; F: 9am-5pm; Sa-Su: 1-5pm
Law: 147 Law Building (333-2914) *	M-Th: 8am-midnight; F: 8am-9pm; Sa: 10am-9pm; Su: 10am-midnight; Sun 9/2: 10am-9pm; Mon 9/3: 10am-midnight
Literatures and Languages: 225 Library (333-2220)	M-Th: 9am-7pm; F: 9am-5pm; Sa & Su: 1-5pm
Map Library: 418 Library (333-0827)	M-F: 8:30am-5pm

Figure 9.5. The University of Illinois library hours table.

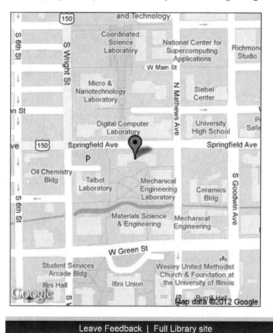

Figure 9.6.   Find a library.

undergraduate or main library reference desks, and users who click our "Text a Librarian" feature on a capable handset will launch a text-message session with a chat reference operator.

We have received valuable input from a "Give Feedback" link we included in the footer of the mobile site: A user can submit comments or criticisms from any point in the mobile interaction. Feedback has been largely positive; we also received requests for additional services, including enhanced access to journal literature and more OPAC features. Such feedback led us to develop access to the "My Account" feature of the VuFind catalog, which allows users to manage their requests and renew materials. We also incorporated the course reserve functions of the OPAC; a frequent request at public service desks at Illinois is about the status of in-demand reserve material.

To provide a space for user access to features still in development, we added a "Mobile Labs" link to the top level of the mobile site. This is where we park services when they are not quite finished or if we are still gathering evidence to whether users are particularly interested in them. In the case of the "My Account" and course reserve features, our web transaction logs revealed these to be two of the site's most popular features; this led us to elevate these features to the top-level mobile page.

INTEGRATING WITH OTHER MOBILE SERVICES

We sought to add value to the library's mobile site through an experimental service that would deliver real-time bus schedule information to students and faculty on the campus. This occurred to us primarily because the Champaign-Urbana Mass Transit District had recently released a very straightforward application programming interface that allowed our software to query the wait times for campus bus stops and receive an XML response. While this service may seem to fall well outside the library's central mission, we included it because it cost nearly nothing to develop or maintain (after the initial XSL style sheet was written, there was little cost to maintain the service—it is administered by an outside authority) and because we reasoned that it would be a useful marketing tool. Students and faculty may not necessarily need to use the library's mobile website every day; however, we felt that the bus stop feature was of particular value to mobile users, and we reasoned that if we provided it, users would be more likely to recall the site's other features when they needed them (Figure 9.7).

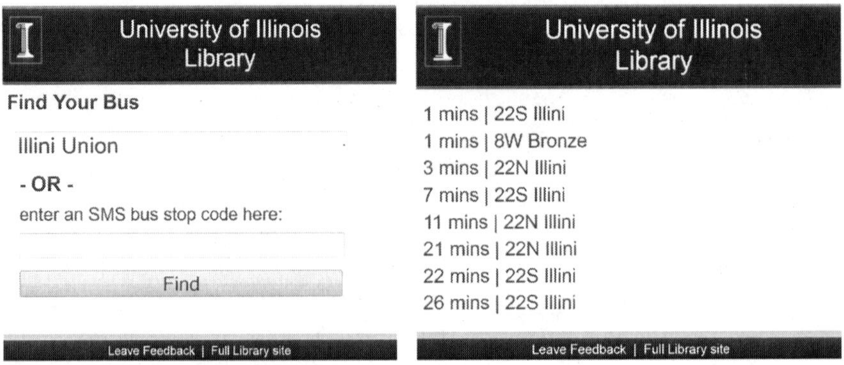

Figure 9.7.   Bus stop information.

Another important consideration in developing mobile services for the library was to communicate with other web services offices within the university. Since fall 2009, the University has advanced various iterations of its campuswide mobile website and has developed full-featured mobile apps for the Android and iPhone platforms. The library has always made certain to communicate with these campus-level web services about their mobile initiatives, and we've successfully advocated for the inclusion and promotion of the library's mobile site in the development of universitywide mobile device services.

LAUNCH, RESPONSE, AND ASSESSMENT

The public launch of the library's mobile website in March 2010 was coordinated by the Mobile Services Working Group and the library's assistant director for advancement for publications and public affairs. A large component of our marketing plan for the public launch involved purchasing print advertising on campus buses; we suspected that a bus ride was an opportune time to introduce a mobile device service to our users. Our advertisements contained screenshots of the mobile site on a simulated phone and directed users to the mobile URL, http://m.library.illinois.edu. We also printed up a series of attractive miniature business cards with the mobile URL and distributed them to users during instruction and orientation sessions, during library events, and at public service desks.

The library's assistant director for advancement distributed a press release to local, state, and academic/higher education media, introducing the features of the mobile website. The launch was also publicized over library and universitywide Twitter and Facebook channels. Several librarians were interviewed for an article in the student newspaper.

To develop staff familiarity with the service, we scheduled two voluntary show-and-tell sessions to introduce the features of the mobile site. We anticipated that not all library staff would have smartphones, so we borrowed several iPod touch devices from our undergraduate library so that staff could experience the site on a mobile device.

Our primary method of assessing the impact of the service has been through analysis of the host server's web transaction logs. The mobile website has consistently received an average of 1,500 unique visitors a month since its launch (there is a significant spike in usage during finals each semester, and there is a drop in usage during breaks and summer months). The most popular features of the site are ranked as follows (based on data from fiscal year 2011–2012):

1. Library Catalog
2. Find a Library

3. Mobile Databases
4. My Account
5. Mobile Labs (contains experimental features, including bus stop data)
6. Course Reserves

After several months of consistent evidence from web logs that users were adopting the mobile website, the site was taken off beta status and adopted as a production service with a maintenance commitment from the library's information technology department. The library requested and received funding from the campus to support continued mobile development: This allowed one of our visiting librarian software developers to split his time between enhancing the mobile site and supporting the grant projects that had initially funded him.

COSTS AND ORGANIZATIONAL CONCERNS

The costs of implementing the mobile website are somewhat difficult to estimate. The virtual server that hosts the server is only a tiny part of the library's web infrastructure, and no new equipment was necessary to deploy the site.

As with many emergent projects, the largest cost is the faculty and staff time that was devoted to discussing and designing the site, gathering feedback, and developing the underlying software. The University of Illinois Library has a policy that allows and encourages librarians and academic professionals to spend a portion of their work time pursuing projects not necessarily related to specific work responsibilities; this investigation time was essential to the success of this project. As the needs of library users evolve, it is crucial that libraries commit staff resources to the exploration of new services and remain responsive to new modes of user engagement.

NEXT STEPS

The most frequently requested new feature for the mobile site is enhanced access to journal literature and abstracting and indexing services. Since this content is primarily hosted by third-party vendors, it has been important for us to advocate to vendor representatives that our users ask us for mobile access to their content. Several publishers and services have developed mobile interfaces, and we have made it a priority to keep up on these developments and add links to vendors' mobile content as it becomes available.

Other mobile features that we hope to launch soon are automatic device detection and a more mobile-friendly proxy authentication page. Several users have mentioned that our proxy log-in page is cumbersome to use on a mobile phone, and we have explored developing a version of the page optimized for mobile devices. Another feature, mobile device detection, is a method by which we program the server for the library's main web gateway to recognize whether a visitor is using a mobile device; if the device is mobile, the program intervenes and asks if the user would like to be directed to the mobile library site instead of the full desktop site. We have not yet implemented this, out of concern that a large volume of device detection may adversely affect our server performance; our testing of the feature has not yet convinced us that our approach scales to handle the several hundred thousand hits that the main web gateway handles daily.

Recently, several librarians from the undergraduate library secured funding from the Institute for Museum and Library Services for a project to codesign library mobile services with groups of undergraduate students. The processes and outcomes of this project will certainly inform additional features of the library's mobile website.

DISCUSSION AND ADVICE

With the development of library services, it is useful to approach them from a view that mobile services do not need to completely replicate all the functions of a library website. Mobile library services are a complement to the variety of web-based services a library offers. With this in mind, it becomes easier to develop a list of functional objectives for a mobile site.

We have found it a very useful practice to include a "mobile labs" space on the developing mobile website. Our lab space gets a relatively large amount of traffic, and it is important to have some way to highlight features or programs in various stages: The feedback that users have submitted about our lab features has been invaluable. Such a space offers developing ideas an opportunity for exposure and comment without serious penalty for failure.

It is crucial to keep mobile development as simple and straightforward as possible. The technical landscape of mobile devices will continue to diverge, and new devices will continue to proliferate. It is important to develop solutions that are independent of the success or failure of different devices; to reiterate, application development for specific mobile operating systems is likely a mistake. Moreover, software development in libraries has a traditional strength in server-side scripting and dynamic website development, and it would be costly to develop new expertise rather than exploit already well-honed skills. If your library is good at making websites but has no experience making apps, then it might make sense to think of mobile web services as nothing more complicated than "little" websites.

The most crucial takeaway from the process of mobile development at Illinois is that most of our challenges were not technical. Our most difficult task surrounded the simplification of an overly complex website; the constraints of mobile presentation forced us to make hard decisions about our information architecture and which features were most important to the goals of a mobile user. On a similar note, a useful outcome of this process has been its influence across all our web services. As we examined the failures of different pages on mobile devices, we discovered ways in which our desktop web presence might be overly cumbersome for users. As we developed enhanced services for mobile devices, we realized areas of our full-sized webpages that could benefit from them. Our mobile initiatives have brought about a healthy reexamination of the usability of many of our services, and now, as the library explores new services, such as web-scale discovery, next-generation catalogs, and growing e-book content, we have a concerned, well-developed awareness of where to position these services for our increasingly mobile users (Figure 9.8).

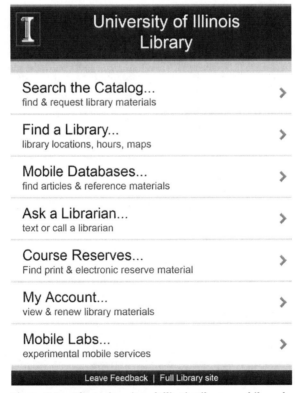

Figure 9.8.   The University of Illinois Library mobile website, http://m.library.illinois.edu, August 2012.

NOTES

1. Aaron Smith, "Nearly Half of American Adults Are Smartphone Owners," Pew Internet & American Life Project, March 1, 2012, http://pewinternet.org/Reports/2012/Smartphone-Update-2012/Findings.aspx.

2. This process adapted from Google's *Static Maps API V2 Developer Guide*, http://developers.google.com/maps/documentation/staticmaps.

10

Building the Montana State University Library Mobile Web App with the jQuery Mobile Framework

Jason A. Clark
Montana State University Library

As the mobile platform was rising in 2010, Montana State University (MSU) Library was looking to introduce services and pieces of its library collections that might make sense in a mobile setting. Talk of a custom iPhone app began, but the web development group was struggling with the notion of learning and maintaining a code base that worked on only a single proprietary platform. There was also interest in building a mobile web app that could work across many mobile platforms. And our commitment to the land grant mission of our university, which promises education accessible for all, only reinforced the need to provide library services and collections using an open platform. As the discussion continued, the idea of a hybrid mobile web app started to gain traction. Our goal was to build a web-based app that answered the needs of our mobile patrons. By applying our existing web skills—namely, HTML, CSS, and JavaScript—we thought that we could produce a simple web app that would work in a number of mobile settings (tablets, smartphones, e-book readers, etc.). The result of this project work was the MSU Library mobile web app, available at http://www.lib.montana.edu/m/ (Figure 10.1).

As of this writing, the mobile web app consists of six primary sections of library content: Search, Databases, Hours, Ask a Librarian, About, and Where. The Search view utilizes the MSU Library "Summon" discovery layer search from Serial Solutions to allow people to search books, journals, and articles in a mobile setting. The Databases page view links into a number of our subscription databases that have been selected from our subscription inventory tool and reworked using our local application programming interface for display. The other page views are what you might expect: Hours, a list of our building hours; Help, links to our reference librarians and our

123

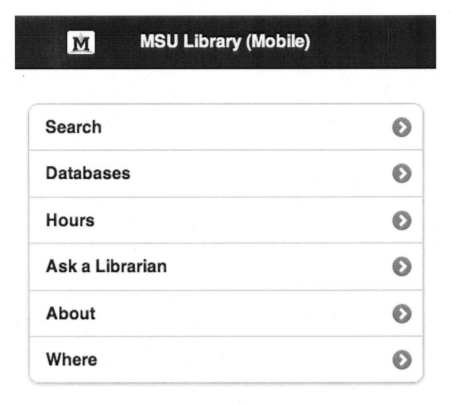

Figure 10.1. Home page view of Montana State University Library mobile web app.

online chat; About, a news feed and contact information; and Where, a Google Map with locations and directions for the library.

Regarding the idea for the web app, our analytics at the time were showing a small percentage of our user population accessing the main desktop website via mobile devices (Figure 10.2). We were clearly at the beginning of the curve, with the analytics suggesting an interest, but not a prescribed need, for a mobile presence. Moving forward with limited supporting data proved to be one of the project's first obstacles. Within the web development group, there was strong consensus supporting the development of a mobile presence and a recognition that the mobile platform was about to become an expectation from our users. We could also see from the analytics that there were certain types of library content that people were using in the mobile context (Figure 10.3).

Our ability to mine the data for these mobile content interest points proved to be a way past the first obstacle. Using the mobile content data, we could begin to construct a working list of mobile page views that would need to appear in our prototype. Top content that we were seeing in this first evaluation included hours, directions, staff information, contact, and an interest in searching the basic inventory of the library. With this well-defined scope in mind, we were able to garner support from interested parties in the library by limiting the development to a focused set of tasks and actions aimed at the library user "on the go." We could build quickly and work without an extended vetting or organizational oversight process. Prior to this mobile project, the Digital Access and Web Services team had built a practice space for applying new web technologies and demoing experimental apps. Our early mobile web app prototype could be made public and tested immediately on this beta section of the website (http://www.lib.montana.edu/beta). This space proved to be another way forward, as it allowed us to debut the mobile prototype and give it an early home without a deep commitment from the organization.

CHIEF AIMS AND OBJECTIVES

As the project started to take shape, the web development group sketched out a number of objectives for the mobile web app prototype. Our first objective was to build our app using common web technologies that were widely adopted, supported, and teachable. In addition, the project needed to leverage existing skill sets for maintenance and further development purposes. We accomplished this by utilizing our existing knowledge of HTML, CSS, and JavaScript to build the app. We also made a decision to use the jQuery Mobile web framework, as it would allow us to work with the most widely adopted HTML web framework and utilize the knowledge of an active, and rich developer community. A second objective was a requirement for an app that would reach as many of our mobile patrons as possible. To this end, the mobile web app needed to be accessible on the network and work across multiple platforms and devices. The first step here was the decision to move away from the native app development platforms of iOS and Android and into mobile web app development. Furthermore, we targeted the most popular devices—namely, smartphones and tablets using Android

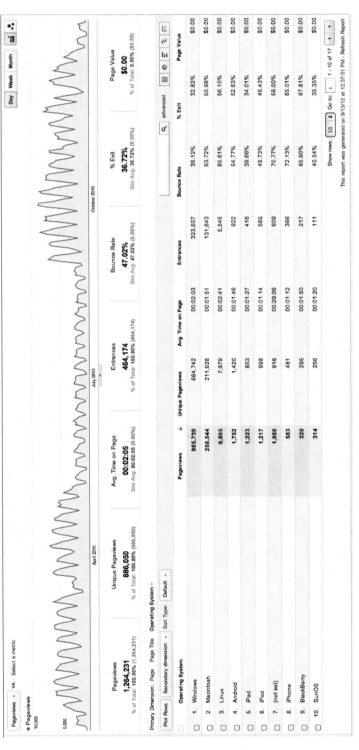

Figure 10.2. Google Analytics showing mobile device use in 2010.

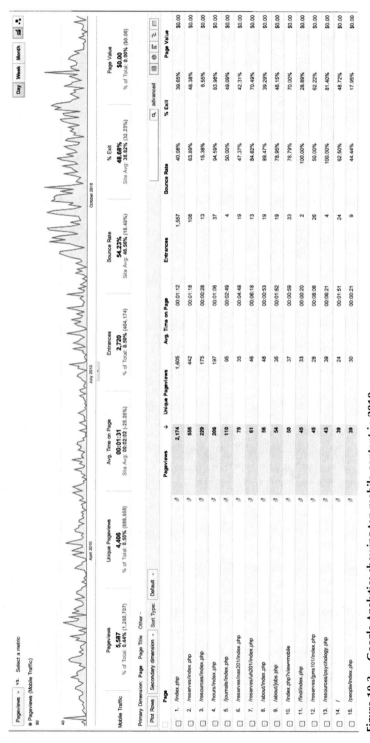

Figure 10.3. Google Analytics showing top mobile content in 2010.

and iOS operating systems—to try to maximize the accessibility of our mobile web app. A third and final objective included a requirement that the app work to surface library content that made sense in the mobile environment. Early on, the group determined that its mobile web app needed to answer the fundamental needs of mobile users, which we identified at the time as location-based information, contact information, and rudimentary searching of library materials. We used our analytics data to guide our prioritization of these content types, but this was probably one of the hardest parts of the project. Selecting the top six core actions of the current site and how they might apply in the mobile setting meant that certain popular sections of library content would not be represented.

IMPLEMENTATION

We began with a soft rollout in the first quarter of 2011. At the time, the decision was made to provide a link in the footer of our desktop website as the primary means of access. In hindsight, this indirect method was not ideal, as it required our users to know that we had a mobile web app and that the app was linked in the footer of our pages. Initial exposure to the app was limited because of this decision. Our logic at the time centered on giving the user an option to select the mobile web app version of our site, rather than forcing the mobile presence on someone. This method of allowing people to "opt in" to mobile presence was an industry standard at the time, but as our mobile user population grew, there came a tipping point where people were expecting to see a mobile view when visiting from a mobile device (more on how we learned and reworked this part of our mobile implementation in the assessment section of the chapter).

Early on, the web development group had surveyed a number of the mobile web frameworks to determine if one made sense to use for the project. In creating mobile web apps, these frameworks do much of the heavy lifting for you and create mobile applike interactions for many of the mobile platforms. In our case, iUI, jQtouch, and jQuery Mobile were the specific frameworks that were considered. A final decision was made to use jQuery Mobile for a number of reasons—chief among them:

- jQuery, a JavaScript library, was the most popular and widely used Java-Script framework.
- jQuery Mobile had extensive developer documentation, which allowed us to learn quickly.

- jQuery Mobile had been tested and vetted on just about every mobile platform and operating system, including Android and iOS (our two target platforms).

Because jQuery Mobile was dependent on common web technologies—HTML, CSS, and JavaScript—training for the developer of the app was not needed. We did use a number of our library-owned mobile devices (a Kindle, an ASUS tablet, an iPad, and an iPod Touch that had been purchased for library staff to use and learn about mobile) to test the mobile web app for optimization and performance on various networks and to verify that our design would display on our targeted mobile platforms. Other staff training and user education were minimal, as the only specialized action required of a user was to be able to navigate to a URL in the mobile browser. This was another advantage of our decision to build a mobile web app; it worked with the natural actions and familiar conventions of desktop web browsing.

RESOURCES REQUIRED

Project staff was limited to the web development group within the library. There were six members of the group, including a computer specialist, the digital initiatives librarian, two reference librarians, and a collection development librarian. The project lead and core developer/programmer on the project was the digital initiatives librarian. The other members of the group worked on content analysis, content selection, and testing for the mobile web app. Outside funding for the project was not needed. As the project evolved into an initiative for the web development group, the only real set of funds in play were the salaries of the group members. Costs were also lowered because we were not outsourcing app development or asking for additional devices and equipment for the project.

On larger, more complex projects like this one, the web development group drafts a project blueprint to define the outcomes, goals, and time-lines for the project work. A template project blueprint is available at http://goo.gl/E4017 and is largely based on Project One-Pager, the project management model created by Tito Sierra of MIT Libraries (http://www.slideshare.net/tsierra/the-projectonepager). For the mobile web app project, we estimated about 3 months from start of content analysis to production and release of the mobile web app. Work began in late December 2010, and the release of our beta app occurred in early April 2011. This was a fairly quick turnaround enabled by our reliance on analytical data to determine content needs. If live user testing with the app had been one of the project milestones, we might have needed another couple of weeks to run the testing and then iterate our design based on the findings.

USER REACTION AND ASSESSMENT

With the soft rollout in April 2011 and limited visibility of the link into the app, initial responses were muted at best. We did receive a handful of comments from our website feedback form thanking us for the mobile version of our site, but much of the MSU Library community was not aware of the rollout. This was one of our larger mistakes in the initial phase of the project.

With respect to furthering our library and university mission, I mentioned the essential role of the land grant mission, which was complemented by our mobile web app efforts. A mission of MSU is to make education accessible for all, and by proxy, the MSU Library adapts this mission by extending the line of thought to include information being accessible to all. The move to create a mobile presence for the library's primary content answers this mission, as it allows for anyone with a mobile device to browse and use select pieces of the library website. In addition, our commitment to developing with the mobile web app approach and its emphasis on open web browsing and networks ensures that the content we provide is not hidden away on a specific platform or device. And finally, part of the charge of the web development group is to experiment and find new ways to apply emerging technologies and platforms. The mobile web app project extended this charge and gave us all a chance to put one of these emerging technologies and platforms into practice.

In our assessment on the mobile web app, we have focused on the web analytics data to help provide a picture of how the app is received and performing. One of the first steps we took was to find a way to capture the analytics data for each page view within the app. A quick aside: one of the benefits of jQuery Mobile is that you build all of your app into a single HTML file and the JavaScript pushes a selected snippet of that HTML into the browser view when a user touches/clicks a specific link on the page. It is a simple technique, performs well in all mobile settings, and makes for an easy learning curve, but it does not play well with Google Analytics. Google Analytics can record only the single request for the full HTML file, not the snippet of HTML that is requested. As a result, your initial analytics data from Google show only one page as being accessed. However, there is a simple fix for this, and the web development group modified the Google Analytics code to allow it to record each snippet of HTML that was viewed. To see the how Google

Analytics can be modified and added to jQuery Mobile, visit the code sample at http://goo.gl/GhSs6 and take note of lines 11–23 and lines 54–63. With the Google Analytics working to record the page views, the web development group started to monitor use and functionality of the mobile web app. Our first 8 months of data showed a steady increase in traffic, especially as the fall 2011 classes started (Figure 10.4).

As members of the web development group worked together and reviewed the data during weekly meetings, it was becoming clear that the app was not being found by all of our mobile users. Our Google Analytics data on the desktop website were showing almost 50% more mobile traffic than the mobile web app, and our "opt-in" and "find the link in the footer" approach was not serving mobile user needs. In January 2012, the digital initiatives librarian began work on a PHP script that would watch for mobile users by matching the user agent string in the HTTP headers of a webpage request to identify specific mobile platforms. Once a user was identified as "mobile," the script would redirect one to the mobile web app as the entry landing page into the MSU Library web presence. This processing of the script was seamless and performed on the server side; the user was simply redirected without a hitch. With this mobile redirect in place, our analytics data have started to reflect our true mobile audience, and members of the web development group continue to monitor the data and what they might mean for the next version of the app.

NEXT STEPS

One thing that is missing in our assessment at this time is strong testimony from our users. As a group, we have tended to rely on the raw analytics data to provide a complete picture of user intentions and actions. While this approach has been useful, it does not give voice to our mobile patrons who might have specific requests and ideas for what the mobile web app should be able to do. With this in mind, an essential next step will be working with a focus group to gather representative opinions. One of the ways that we can gather this information is through our library curriculum courses. We also plan to conduct some low-impact user testing where students and faculty are invited to use the app for a set of prescribed tasks and a single member of the web development group will observe their satisfaction and success rates.

Even without this user testing, we are learning from our analytics data about core content and actions that people are doing within the app. Specifically, the web development group has noted an interest in more robust searching and the need to perform common library research actions, such as

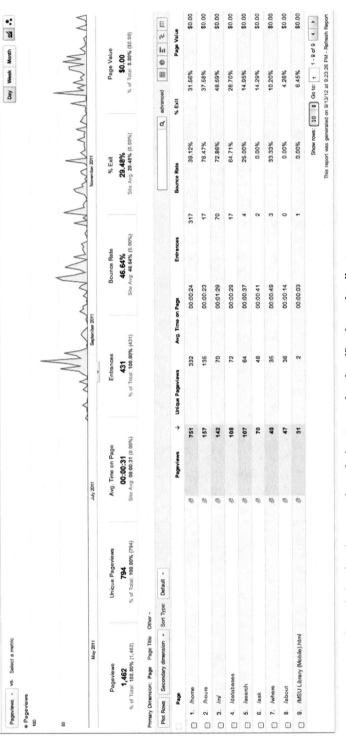

Figure 10.4. Google Analytics showing page views during 8 months of traffic after soft rollout.

checking library accounts, requesting items, and retrieving reserve materials. The expectations for mobile seem to be shifting away from simple "on the go" actions, such as checking hours, to the more familiar library research actions, such as downloading a reserves item. Given these rising interests, the next iteration of the mobile web app will need to include or address these trending interests. To this end, a "library accounts" page view along with a "library reserves" section seem like logical next steps. One of the challenges in bringing in these new sections will be balancing the simplicity of the current app with only six browse points and the need for more complex, nuanced interactions that will require multiple screens and data views.

The web development group has also been interested in the possibilities that translating the current mobile web app into a native application would create. One of the driving interests here has been the ability to market the app in the Apple App Store or the Google Play Android marketplace—not necessarily in the interest of monetizing the app but more in the interest of establishing the MSU Library as an active and reliable developer in an emerging technological market. New mobile software frameworks, such as PhoneGap (http://phonegap.com), can take your mobile web app and convert it into a native application that can be deployed in app stores. The web development group at MSU Library is considering PhoneGap, as it would allow us to build a mobile web app that would be converted to a native Android and iOS app without a hitch in our workflow. The group is also considering HTML5 frameworks such as Sencha Touch (http://www.sencha.com/products/touch/), which can create native applike experiences by using cutting-edge HTML, CSS3, and advanced JavaScript techniques. Plans for each of these potential initiatives are in early stages, and a fuller discussion of developer and staffing resources as well as web development team priorities will guide future work in this area.

ADVICE FOR OTHER LIBRARIES SEEKING TO DO THIS

If you are getting started and thinking about how you might make the move to a mobile presence for your library, there are a number of takeaways and lessons learned that our group can share.

First: Have a Promotional and Marketing Plan for Your Mobile Project

Our "soft rollout" approach limited the uptake and interest in our app. We were able to recover, but there was no need to be in a position where our app was not findable by all our mobile users. Our project blueprint could have included a marketing plan bullet. The plan might have been as simple as requiring a blog post about the project, creating a talking-points handout for library staff to allow for advocacy and promotion of the app, and building a simple redirect script that would place the app in front of the majority of our mobile users.

Second: Remember That Data Are Your Best Friends

You can see how we used our analytics data to guide the building of our prototype mobile web app. Initially, our web analytics data set informed the selection of content for the prototype by highlighting those pieces of the desktop site that our mobile patrons were using. We also used the web analytics data set to analyze how our mobile users were not finding our app, and that led to an essential fix (the mobile redirect script) that brought a true audience to the app.

Third: Build Early and Build Often

The web development group made a conscious effort to sketch out ideas for the app and then quickly move those ideas into a working HTML demo. If a picture is worth a thousand words, prototypes are worth at least two thousand. They fix perceptions about what is possible and keep groups focused on how the tool you are building needs to work. Hypothetical interactions and questions of language labels, which can derail and stall discussions, fall away when you have a demo in front of a group. The conversation tends to focus on how to solve the problems that the group is seeing in the demo. Use that to your advantage, and know that your prototype is a draft that can be reworked.

Fourth: Start Simple and Enjoy the Liberation That Mobile Design and Development Enables

With limited screen sizes and the need for larger touch interface inputs, mobile design asks you to be smart about arrangement and white space in your app. This is not a bad thing. Mobile design limits what you might put on a page view, but these limits can be freeing, as they really ask you to focus and prioritize the core actions and content for your app. Some of the best discussion of our group had centered on boiling down the library desktop website to six primary actions. On a related note, mobile development requires you to be smart about optimizing images and files, caching of files, writing clean HTML and JavaScript, and limiting HTTP requests to allow your app to perform well on the limited bandwidth of mobile networks. Building for mobile has limits, but they can improve your design and development skills.

Fifth: Paying Attention to Emerging Development Techniques and Technologies

Responsive web design (http://en.wikipedia.org/wiki/Responsive_Web_Design) is a web development technique where a foundational HTML markup is built to work in all kinds of different containers based on screen sizes. If you think about how water works to adapt to whatever container it is poured into, you start to get a mental image of the concept of responsive web design. The secret sauce is a set of rules in the CSS file, the media query, that create uniquely formatted displays based on how large or small the browser window is. HTML5 is an umbrella term for a suite of technologies centering on new HTML markup, advanced CSS features, and using JavaScript to access and interact with application programming interfaces, which can be used to build web apps that run in a browser but feel like native applications. In our case, the jQuery Mobile framework uses HTML5 to demarcate pages in the app and media queries to help create the default mobile display for our app. Knowing what these techniques and technologies enable is a step in learning where mobile development will be going in the future.

Sixth: Finally, a Reminder to Not Reinvent the Wheel

When our web development group began its research into the possibilities for our app, we knew that we needed to invest in a code resource that would work across most, if not all, mobile platforms but also be maintainable by any number of developers that might work after we had moved on (or retired). jQuery Mobile, with its widespread adoption and use of common web technologies, had all of these qualities and was intuitive enough to allow for a short learning curve. There was no need to start from scratch, and that allowed us to extend our app even further than we anticipated.

In the interest of helping others even further, I have built a mobile library web app starter kit using the jQuery Mobile framework. This kit is the simpler version of our current MSU Library mobile web app, which is perfect for learning the basics of jQuery Mobile and modifying it for your own needs. You can view the demo here: http://www.lib.montana.edu/~jason/files/touch-jquery/. If you are interested in downloading it, grab the zip file at http://www.lib.montana.edu/~jason/files/touch-jquery.zip.

11

The Gimme Engine: A True Story of Innovation, Creativity, and Fun

Aimee Fifarek and Ann Porter
Scottsdale Public Library

In 2010, we at the Scottsdale Public Library decided that the time had come to build our first mobile web app. We had been following the "be where your customers are" philosophy for some time, including outreach to community organizations, taking our early literacy programs to day cares and community centers, and gearing our website increasingly for home use. However, given the incredible increase in technology, we knew that there was one area that we had not yet reached: our customers' smartphones.

So we created the Gimme Engine: a random book recommendation app optimized for mobile devices (Figures 11.1 and 11.2). Go to http://gimme. scottsdalelibrary.org, choose a category that sounds interesting to you, and click "Go." "Gimme a Clue" will recommend a mystery novel; "Gimme Some Lovin'" will recommend a romance; "Gimme Liberty or Gimme Death" will suggest something historical. Each recommendation gives you basic bibliographic information, the book jacket, a summary of the book, and a review from a Scottsdale Public Library staff member. There are quick links to check the availability of the item, place a reserve, and share what you've found with your friends via Twitter or Facebook. If the first recommendation wasn't what you wanted, just tap "Gimme Another," and you will get a new recommendation.

As awesome as Gimme is, this was not the app we set out to build. The idea of creating a mobile app grew out of a leadership seminar in which some of our younger staff participated in 2009. After about a dozen lectures and as many workshops and small group discussions, each team in the seminar was expected to produce a project in the form of a Library Services and Technology Act grant application. The Scottsdale team developed the "Library Anywhere" concept—a mobile version of the library's website. At the time,

Gimme!
Powered by SPL

Getting a recommendation is easy! Just choose a category from the dropdown menu and click "Go!"

Figure 11.1.   Gimme Engine home screen.

GIMME...

| a clue | ▾ |

GO! ▶

🐦 Tweet   📘 Recommend 69 💬 Send

ₗₗ Verizon 3G   **3:27 PM**   88% 🔋

GIMME...

| a thought balloon | ▾ |

GO! ▶

Angel: After the Fall, Volume 2: First Night

Joss Whedon, Brian Lynch, John Byrne, Tim Kane, Stephen Mooney, David Messina, Nick Runge

9781600103933

Figure 11.2.   Gimme Engine result screen.

Description

In a story that follows the events of the show's final televised season, a heroic vampire seeks redemption while dealing with the death destruction brought on by his choice to stand up to a demonic multi-dimensional law firm.

Review from Scottsdale Library Staff

A postscript to the TV series, this set of graphic novels carries on the adventures of the vampire detective, Angel, and his friends and foes. This particular volume picks up exactly where the cliffhanger left off: L.A.'s descent into hell. A must-read for fans of the show. - Hillary D.

RESERVE ▶   **GIMME ANOTHER** ▶

Table 11.1. Scottsdale Public Library Mobile Web: Library Services
and Technology Act Grant Budget

Category	Amount Awarded	Amount Spent
Contractual services	$12,000	$15,036
Hardware	$2,750	$1,948
Supplies	$650	$1,616
Software	$3,200	$0

library mobile was still young enough that a mobile site was a great idea. So when the grant application time rolled around, the grant experts took the proposal, tightened it up, and submitted it. Three months later, the application for $18,600 was approved (Table 11.1). But then our excitement quickly turned into anxiety because now it was go time: we actually had to build the app.

GETTING STARTED

The first step was to assemble a team of staff members to help guide us through the process. We invited the "usual suspects"—our tech folks, graphic designer/webmaster—but we also invited staff who would act as advocates for the app once it was created, to help garner support from all staff. We included one of the members of the leadership team that wrote the original grant proposal (who is now a building manager), an energetic and imaginative librarian from the youth staff, and an up-and-coming library assistant in adult services with a technology bent. Aimee and Ann were the project leaders, as technology and marketing are paramount in this type of endeavor. It would be our job to not only harness all of the creative intellectual energy of the group but also ensure that the final product represented the right mix of form and function.

Our first meeting was interesting to say the least. We were all over the map when it came to ideas about what this app should actually do. There was still some support to build a generic mobile website, while some folks thought that it should be more interactive, like a scavenger hunt. Over the next few hours, the ideas flew, including the concept of "Dewey Caching"—like geocaching but hiding things in the stacks to help people get to know the library. One of the tech folks was enamored with the iPhone "bump" technology, where you tap two iPhones together to exchange data. Aimee wasn't really settled on any idea except that the app should be platform agnostic—we shouldn't exclude customers from using our app just because they didn't have the right type of technology. Ann thought that we should definitely not limit ourselves on our brainstorming; however, we needed to get a better gauge of what people would actually use.

Clearly, we needed focus. Fortunately, we had put a good chunk of money into the grant to hire a consultant to do market research for us. Thanks to Ann's contacts in the online marketing world, we were able to hire Forty, an "experience design agency" that does market research, app design, and development.

We brought in Forty to meet the team—or, rather, the rest of the team. Scottsdale Public Library does a lot of outsourcing. Upward of 95% of our materials come shelf ready, and our integrated library system is a turnkey system. We treat our vendors like remote-based library staff: we bring them in, educate them about our library, clearly lay out our expectations, and then leave them alone so that they can get to work. We find that this not only results in better quality solutions for our staff and customers but puts our minds at ease regarding the partners with whom we work.

In this case, one of the major expectations was that we wanted Forty to not only use current library customers for the market research but reach out to future customers (what we call nonlibrary users) as well. The Scottsdale Library had a number of initiatives in its strategic plan around building our customer base, including increasing usage among 19- to 30-year-olds, so we were hoping that this app would help to attract that hard-to-reach demographic.

We knew that the best way to get this feedback would be via a survey. We typically survey our customers once a year to gauge their satisfaction with library products and services. So we looked to Forty to spearhead this survey initiative. During our early meetings, we worked with them to determine what questions they should ask and how they should ask them, to get good information to help guide our decision.

The brief but detailed survey was distributed on the library's website, Facebook, and Twitter pages, as well as via our monthly e-newsletter, which reaches roughly 70,000 library customers. Forty also distributed the survey to its e-newsletter customer base and Facebook fans. We ended up with around 200 people responding to the survey, which was posted for about 7 days. Of those, 24% never or seldom visited a library, which we felt was a great representation of our "future customers."

As we broke down the survey results regarding what customers wanted out of a library mobile app, we noticed something very interesting. The top five services they wanted were those we already offered through our mobile catalog (http://mcat.scottsdaleaz.gov). This included book renewals, book availability, a list of new items at the library, the calendar of events, and searching the library catalog. The sixth and seventh services they wanted were two that we didn't offer—staff reviews and book suggestions. Because of this, the library team agreed that we needed to create some sort of book recommendation engine, powered by library staff reviews. It would be a great new service that we could offer our customers and one that all staff could rally behind.

We scheduled a meeting with the team from Forty to discuss our takeaways from the survey results. When we got together, the team at Forty was leaning toward creating an app that met more of the items that survey respondents indicated they wanted. We did follow its logic; however, we strongly felt that it was more important to step outside the box a bit for this grant project. We took a week to continue to brainstorm how we could harness all the library staff book reviews and deliver them to customers in a way that was easy, fun, scalable, and sustainable.

During one of our meetings, we acted out a typical exchange between a customer and a staff member when the customer was asking for a book recommendation. Most of the interactions began with "I really like books by Author X. What other books would you recommend?" or "I really enjoy spy novels. What suggestions do you have for something for me to read?" In essence, the request was "Give me a good to book to read."

This is when the brainstorming really got interesting. We knew that we would need categories of books that customers could search, as that would be the easiest way to organize the service. We also knew that the categories couldn't be "vanilla"—they needed to spark some interest to get users engaged.

At some point during one of the more goofy stages of brainstorming, Ann said, "We should just call it 'Gimme,' and then the category could be 'Gimme Something Good to Eat' and the result would be a cookbook. Or 'Gimme Some Lovin'' and we could recommend a romance." She was mostly joking, but once the idea was spoken out loud, the team liked how it sounded. We then began to flesh out more details about how the Gimme concept could work, and before we knew it, we were onto something.

Earlier in the year, the Scottsdale Library had launched a Goodreads account to share book reviews. We had about 80 book reviews on Goodreads at that point, and we figured that we could start with those 80 books to help power the Gimme. Then we would be leveraging the readers' advisory skills of our staff in an innovative way, without creating more work for them. They were already creating these reviews, so why not use them for the mobile app?

Needless to say, we were excited. We did take some time to consider the fact that we were about to move forward with a library service called Gimme—which is not considered perfect English. We are well aware that our customers can be vocal about grammatical errors made on our website and calendar of events, so how would they react to a new service with a slang word for a name? We decided that the spirit of Gimme was to be fun, out of the box, and different, so we would move forward.

DESIGNING A GREAT LIBRARY APP

We set up a meeting with Forty, and it too loved the idea. We then began the process of developing the mobile design for Gimme. We knew we wanted it to be simple (as it would be viewed on a phone screen—no need to go crazy with graphics), easy to use (because our customers have a variety of techni-cal experience), and timeless (since we were working with grant monies, we would not have the funding to go back and do a Gimme face-lift in a few years). Forty came back with some great designs that fit all of our criteria.

We had the concept. We had the design. Now we just had to make it work. Forty brought in a very young, very energetic programmer to develop the Gimme technology. In the one and only face-to-face meeting we had with him, we discussed the Gimme concept and what technologies and resources the library had at its disposal: a mobile catalog (Innovative Interfaces' Air-PAC) but no application programming interface for merging data, the ability to create topical RSS feeds from the catalog (Innovative's FeedBuilder), the library's website using the Expression Engine content management system on a locally hosted server, staff reviews living in Goodreads, existing social media presences on Facebook and Twitter, and a small amount of money to buy software for this project.

After turning this over in his mind, the programmer surprised us both by saying that we wouldn't need to buy any software to make Gimme happen. He would write some JavaScript that would randomly select a title from the RSS feed when a category was chosen—but not too random, to avoid someone getting the same recommendation three times in a row. He would then match up that title with the corresponding staff review by using the free Goodreads application programming interface. The matchpoint was the ISBN number of the book—the one consistent data point that was present in both the catalog and the Goodreads database. He could then link back into the catalog so that customers could check the item status and place a hold, again using the ISBN. We simply had to use some of the software money that we had just saved to have the ISBN included in the keyword index.

While he worked on getting the programming in place, we realized that we needed to work on our internal processes. Aimee needed to figure out how to get specific titles to show up in the RSS feeds, and Ann had to get staff who were writing book reviews to write them specifically about the Gimme titles.

GIMME INFRASTRUCTURE

To date, the library had used RSS feed primarily to show what new items were coming into the collection. It works by using queries that retrieve bib-liographic records based on common datapoints, such as location or subject. The RSS file for a specific feed references that query to get the records and has other settings in the file that tell the system what field in those record should be displayed. The feed automatically includes the newest records that match the query criteria. However, in this case, the criteria were not consistent—specific unrelated titles had to be retrieved. That meant that something unique would have to be added to those records in the database. Only one group would have the solution to this one—our crack cataloging team.

Aimee met with the catalogers and explained the idea of the RSS feeds and the problem of retrieval. They came up with the idea of using the phrase "Library Staff Review" in an 830 field, followed by some piece of controlled vocabulary that would describe the Gimme category—for example, "Library Staff Review Mystery" or "Library Staff Review Romance." Since the series statement was indexed, it would allow staff and customers to use these phrases in the catalog to retrieve records with reviews as well—something that our discovery product does not do through the existing facets and in-catalog review process.

That was great for catalog retrieval but not quite unique enough for the query. It was unlikely that other items would get retrieved by those searches, but it was possible. That's when we hit on the idea of punctuation. If the encoding was "Library Staff Review_romance" it would be possible to set the query to retrieve only records with "Review_romance" in the 830 field. Since that structure was a nonstandard cataloging practice, it would truly be unique, yet it would not affect the use of the statement for staff and public because the discovery platform ignores all punctuation. We set up a test query and feed, and it worked beautifully. One very big problem was resolved.

However, now that the catalogers were going to need to touch every MARC record that was reviewed they needed to know which records those were. Up until this point, the staff doing book reviews were a loosely knit group who were simply adding reviews to Goodreads whenever they read a good book or received a review from another staff member. Ann had given them some basic criteria about reviewing—basically, that they should be reviewing only good books that they wanted to recommend; other than that, it was a somewhat ad hoc process.

So we met with the Goodreads group, at that point about 10 staff, to tell them about the Gimme project and get their ideas for formalizing the process. We wanted to be careful and not make what they were doing all about Gimme, but we wanted to enlist their help in making it a success.

The first idea that came up was to put something up on SharePoint, which had recently replaced our old Intranet system and was our shiny new toy. Everyone loved it, so it made perfect sense to create a review submission form there. The Goodreads group agreed to do an advertising campaign to get more staff to submit reviews and prioritize posting the reviews that matched the Gimme categories.

The Goodreads group also created a workflow in SharePoint to manage the reviews and interact with the cataloging team. All the reviews submitted went into a spreadsheet-like interface. They created a calendar on SharePoint, and each person volunteered to take certain days where they would check for new reviews and post them to Goodreads. They would mark the review they did as closed, which would be the catalogers' prompt to add the series statement to the MARC record. The catalogers decided that they would also add the text of the review itself to the MARC record so that we could be sure to retain our data, should we ever have to stop using Goodreads. The catalogers would then delete the record from the review database, and the process would be complete.

We started to build up the feeds. It was working fine for the most part, but as the programmer started sending us test versions of the app, we noticed that reviews were missing from some of the records and some more serious errors were occurring. It didn't take us long to discover that the problem was that we were sometimes adding our data to different versions of the same title. The ISBNs weren't matching.

As it turns out, there are a large number of titles on Goodreads that list multiple ISBNs—all legitimate editions, just different ones. Our staff were simply choosing the book in Goodreads and posting the review. To resolve this problem, we needed to make sure that the review in Goodreads was attached to the book with the same ISBN as the one in our collection.

We also needed to make sure that we had sufficient copies of the reviewed titles in our collection to avoid creating long hold lists and delaying Gimme customer satisfaction. The catalogers had no problem incorporating this item check into their process. They were already choosing the record with the most available copies of the title for the review and the encoding. We then followed up with the Goodreads group to let them know the importance of the ISBN in this process and ask them to check the catalog for item availability prior to posting the review on Goodreads.

The last part of the building process was for the programmer to provide an administrative interface. He made it very simple, which was just what we wanted: one box for a title, one box for the URL for the feed, and a submit button. Once a new feed is submitted, its entire contents appear below the form so that you can check for problems with entries. Each feed has an "update" button and a "delete" button, so we can easily bring in the newest entries. With that, Gimme was ready for prime time!

GETTING READY FOR GIMME

The development of Gimme was being done in a very public way. Between the Gimme Brain Trust and the Goodreads group, we had at least 20 staff directly involved, not to mention the number of staff who had responded to our request for additional reviews. But we wanted to make sure that every member of staff who might need to talk to customers about Gimme was familiar with the app and understood the basics of how it would work. Due to budget cuts, the library no longer had the luxury of sending staff to regular training sessions. Most training is carried out at meetings when staff is already scheduled to be away from public service points.

With grant funds, we had purchased five iPod Touch devices. We thought it would be the best way to simulate the mobile phone experience without actually having to provide a phone (with a service contract) for staff use. One of the members of the Gimme Brain Trust took on the task of setting up the iPod Touch for each library—adding the on-screen icon for Gimme, installing other mobile apps for the various database and language products we have, and writing up a "best practices" guide for using the mobile device in a public service interaction. One of the members of our Spanish-language outreach team heard about the project and volunteered to write up a best-practices guide for using the device with Spanish-speaking customers as well. We sent out the devices to the libraries and then got on the agendas of their staff meetings to demonstrate Gimme. This way, staff understood the product before we started our marketing campaign.

SPREADING THE WORD

After extensive testing and training, we were ready to begin marketing it to our customers. Our first concern was finding a way to help Gimme stand out among all the noise. Our library, like most, offers a vast variety of programs and services for customers aged 0 to 100 and from all walks of life. With all the hard work that had gone into the creation of Gimme, it was important to make sure the marketing push would be effective.

One night, Ann realized that the typical marketing efforts wouldn't be innovative enough—though Gimme would be featured on the website, marketed through Facebook and Twitter, and announced to the press via a press release, we definitely needed to do more to ensure that the community at large was aware of this unique new service. Then a light bulb went off over Ann's head—why not leverage library staff to market Gimme?

Seeing as how it was their readers' advisory expertise that customers were looking for, it made sense to use them in the marketing efforts to reinforce the value of that service. To bring the staff recommendations to life, Ann decided that it would be best to ask staff who had contributed reviews to have a life-sized cardboard cutout of themselves created. She asked them to bring props for the photo shoot that best represented who they were as a person—to show off their personality. Ann sent an e-mail out to gauge interest, and 15 brave souls volunteered.

Library staff brought the most interesting props—cameras, aprons, kids, dogs, quilts, microphones, bicycles . . . you name it. They really embraced showing of the "who" behind the staff reviews. Once the cardboard cutouts were printed, Ann created speech bubbles to hang above each one, such as "See what she cooks up for a good read by visiting gimme.scottsdalelibrary. org" or "What does this busy mom on the go recommend you read today? Visit gimme.scottsdalelibrary.org to learn more." We also put the library staffer's first name and occupation underneath the statement. The cutouts were a hit with staff and customers alike. They were definitely eye-catching and great conversation starters. They were hard to ignore—and that was the point (Figure 11.3).

We also created 36-inch-diameter floor decals for each branch. These were placed by the entrances and exits so that any customer walking in the door would be greeted by the Gimme logo and web address. We created buttons that mirrored the floor decals for staff to wear, as well as T-shirts for those who wanted to wear them. The point was to have Gimme be front and center when customers entered the library. It also helped our staff to be familiar with our new service, as they were walking advertisements for Gimme (Figure 11.4).

Figure 11.3. Scottsdale library staff: Brad Morse and Ann Porter pose with staff cutouts.

Figure 11.4. Examples of the Gimme floor decals and buttons used to increase awareness.

LESSONS LEARNED

All in all, Gimme was a very successful project. We developed a great app with staff support that capitalized on library strengths; we created new internal processes that could be woven into the day-to-day workload of staff with minimal effort; and we concluded the project on time and on budget. But then one day, someone asked us, "What impact has Gimme had on your circulation?" The only response we could muster was "Um . . ."

When you are working on a project like this, it is very easy to think only of the short-term goal: getting the thing up and running. But you also need to ask the big picture questions: What will be my measure of success long term? Do we have the resources—human, monetary, and technological—to support the end result? Are we prepared to pull the plug when it has outlived its usefulness or become too resource intensive to support?

This is where we are right now with Gimme. Our internal processes ensure that new content will continue to support it. One thing we know for sure is that library staff are not going to stop writing book reviews any time soon. But how frequently do we need to change categories to keep it interesting for both current and new customers? When we add new categories, should we take old ones off (at least for a while)? Should we put efforts into creating seasonal or "hot topic" categories? Creating the category itself is easy, but you might not have enough reviews on hand to create a "*Hunger Games* Read Alike" category when your hold list for *Mocking Jay* is a mile long.

Another key factor that we did not account for is statistics. Almost as an afterthought, we slapped the Google Analytics code up on the Gimme pages but did not have a plan to say what specifically we wanted to report on or, for that matter, develop any statistically relevant goals. Being able to say that we started out at X and now we're at Y is always a good thing, even if (like Gimme) it won't require any additional monetary resources to support.

On the technological front, we are in pretty good shape. The code that was created is fairly simple, so the risk of it having a major break is low. Should that happen, we have a little bit of expertise on staff that will likely be sufficient to handle the problem. What is more likely to happen is that 2 or 3 years from now, the speed of technological change may make the entire Gimme concept seem dated or irrelevant. At that point, we will have to assess our resources and decide if it is worth creating Gimme 2.0 or if the better option is to lay Gimme gently to rest. Then we will have to go through the same process that we do when we consider discontinuing any service: determine how much staff and customers are invested in its existence, and figure out some other way to meet the ongoing needs.

But for now Gimme is in its heyday. The library has earned a number of recognitions and awards, and awareness of our little app is growing. We have already added three new categories: "Gimme a Thought Balloon" (graphic novels), "Gimme a Bedtime Story" (picture books), and "Gimme a Story, Dude!" (teen fiction). At the time of this writing, we have a systemwide initiative to increase usage of e-books among customers by 400%, so we are working on new e-book and e-audio book categories. We will be changing out categories on a quarterly basis starting in January 2013, and our update plan will have seasonal components that match big library events, such as Summer Reading.

As for statistics and measures of success, we are still trying to determine what will be most useful. We extracted a list of all titles in Gimme and can periodically check the increase in their circulation. However, without a set of comparable titles that we can measure them against, it will be hard to prove a direct link between inclusion in Gimme and higher circulation. Probably the best we will be able to do is to note the change in circulation on a monthly basis and compare that to the increase in circulation of the overall collection to see if there is a persistent difference. Is that a meaningful number? It is hard to tell.

In the end, the best measures of success may fall into the intangible category. We created a new service for customers, with no operating budget, that requires very little maintenance. A classic library service that our staff love to do—readers' advisory—now has a 21st-century presence because it is included in a mobile app. And we completed a major award-winning project by incorporating the expertise of staff at a various levels, so we have something to feel really good about. That right there may be the best measure of success.

Index

About the Editors and Contributors

Amanda Binder is the social sciences librarian at the J. Murrey Atkins Library at the University of North Carolina at Charlotte.

Helen Bischoff is the reference and instruction librarian at Georgetown College Library in Georgetown, Kentucky.

Josh Bishoff is the visiting digital library research librarian at the University of Illinois Libraries at Urbana.

Laurie Bridges is an instruction and emerging technologies librarian at Oregon State University in Corvallis.

Anne Burke is the undergraduate instruction and outreach librarian at the North Carolina State University Libraries in Raleigh.

Jason A. Clark is head of digital access and web services at the Montana State University Library in Bozeman.

Matthew Connolly is the Applications Programmer III at the Cornell University Library, Ithaca, New York.

Tony Cosgrave is a Librarian I at the Cornell University Library, Ithaca, New York.

Aimee Fifarek is the library technologies and content senior manager for the Scottsdale Public Library in Arizona.

Alexandra W. Gomes is the associate director for education, information, and technology services at the Himmelfarb Health Sciences Library at George Washington University, Washington, DC.

Charles Harmon is an executive editor for the Rowman & Littlefield Publishing Group. His background includes work in special, public, and school libraries.

Scott La Counte is the author of *Going Mobile: Developing Apps for Your Library Using Basic HTML Programming.*

Adrienne Lai is the libraries fellow in the Special Collections Research Center and User Experience Department at the North Carolina State University Libraries in Raleigh.

Michael Messina is a reference librarian at the State University of New York's Maritime College. He has also worked as a researcher at the Brooklyn Academy of Music Archives. The former publisher of Applause Theatre and Cinema Books/Limelight Editions, he is a coeditor of *Acts of War: Iraq and Afghanistan in Seven Plays.*

Ann Porter is the community relations coordinator at the Scottsdale Public Library in Arizona.

Ben Rawlins is the electronic resources librarian at the Georgetown College Library in Georgetown, Kentucky.

Matthew Reidsma is the web services librarian at the Grand Valley State University Libraries in Allendale, Michigan.

Hannah Gascho Rempel is the science librarian and graduate student services coordinator at Oregon State University in Corvallis.

Adam Rogers is the emerging technology services librarian in the User Experience Department at the North Carolina State University Libraries in Raleigh.

Michele Ruth is the collection development librarian at the Georgetown College Library in Georgetown, Kentucky.

Sarah Sagmoen is the director of learning commons and user services at the Brookens Library at the University of Illinois at Springfield.

Cassandra Shivers is the digital access architect at the Orange County Library System in Orlando, Florida.

Natalie Tagge is the instruction librarian at the Claremont Colleges Library in Claremont, California.

Nancy J. Weichert is a visiting assistant professor and instructional services librarian at the Brookens Library at the University of Illinois at Springfield.

Evviva Weinraub is the director of emerging technologies and services at Oregon State University in Corvallis.